ELECTRIC POWER INDUSTRY

IN NONTECHNICAL LANGUAGE
2ND EDITION

ELECTRIC POWER INDUSTRY

IN NONTECHNICAL LANGUAGE
2ND EDITION

DENISE WARKENTIN-GLENN

Disclaimer

The recommendations, advice, descriptions, and the methods in this book are presented solely for educational purposes. The author and publisher assume no liability whatsoever for any loss or damage that results from the use of any of the material in this book. Use of the material in this book is solely at the risk of the user.

Copyright© 2006 by
PennWell Corporation
1421 South Sheridan Road
Tulsa, Oklahoma 74112-6600 USA

800.752.9764
+1.918.831.9421
sales@pennwell.com
www.pennwellbooks.com
www.pennwell.com

Director: Mary McGee
Managing Editor: Steve Hill
Production / Operations Manager: Traci Huntsman
Production Editor: Amethyst Hensley
Production Manager: Robin Remaley
Book Designer: Amanda Seiders
Cover Designer: Charles Thomas

Library of Congress Cataloging-in-Publication Data

Warkentin-Glenn, Denise.
  Electric power industry in nontechnical language / by Denise Warkentin-Glenn.-- 2nd ed.
    p. cm.
  Includes bibliographical references and index.
  ISBN-13: 978-1-59370-067-6 (hardcover)
  ISBN-10: 1-59370-067-9 (hardcover)
  1. Electric utilities. 2. Electric power production. I. Title.
  HD9685.A2W28 2006
  333.793'2--dc22

                                                                    2006003358

Printed in the United States of America

1 2 3 4 5   10 09 08 07 06

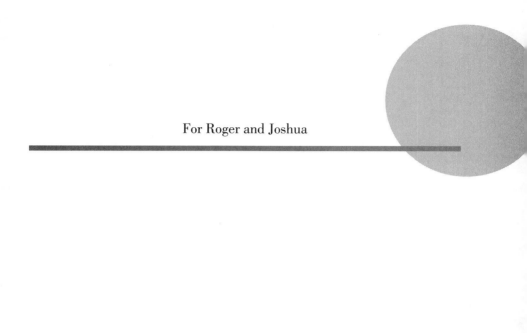

For Roger and Joshua

Contents

Preface

To look at the electric utility industry as a static estate is undeniably a mistake. The many developments going on inside the industry foretell the story of an ever-changing, ever-evolving industry. From what is taking place within individual electric utilities to what is transpiring on state and federal levels, we can learn a lot about where the electric power industry is heading.

What is going on is a plethora of activity germane to how electric utilities operate and will be required to operate—probably far more than what can be covered in this book. The activity swarms around governance on a vast array of issues, such as environmental protection, technology issues and advances, reliability and security enhancements, and generally on the processes of generation and transmission.

As was the case with the 1st edition of this book, *Electric Power Industry in Nontechnical Language,* this 2nd edition will seek to present timely, and oftentimes thought-provoking information about the state of the industry. This book is, simply put, the layman's handbook. The text you are about to read is specifically written for all those non-technical individuals who are wanting to learn the fundamentals of one of the largest and most fascinating industries in the world.

My goal with this book was to provide a somewhat different "read" than the first edition. In the pages of the 2nd edition, readers will be able to get a clearer picture of where the electric utility industry is headed in the future. Thus, the primary focus of this book is more one of how electric utilities are functioning and are expected to function within the current air of restructuring activities. My hope is that readers will enjoy the look back at the early years of the industry and its colorful beginning and evolvement, but will, like me, be most attuned to how electric utilities are going to handle their businesses in the future.

In the following pages, readers will learn everything from the very early beginnings of the electric utility industry to the more complicated realities surrounding the industry today. With you in mind, I have developed the following three parts that may be read consecutively, or if preferred, read out of the order in which they are presented, since each "section" is a concept within itself.

Part I explains the core technical competencies of the electric utility industry. This section touches on generation, efficiencies, fuel, and new technologies affecting such. The reader will also learn about transmission and distribution and, most importantly, how they work together with generation to produce and deliver power to consumers.

Part II examines both the formation and reformation of the electric utility industry. From early regulation to the ongoing movement of industry reregulation, this part will help get readers up to speed on how regulatory strictures began in the formative years of the electric utility industry, how regulation has evolved, and currently how the industry is practically re-inventing itself amidst a climate of restructuring.

Part III discusses emerging issues and trends. Readers will find this section to be a wealth of information on learning about where the electric utility industry is headed in the next few years. Of great importance is the passage of the Energy Policy Act of 2005. Here, readers will find a concise overview of one of the most comprehensive pieces of legislation enacted since 1992. Also covered for general understanding are the issues of transmission, technology, and how they tie into reliability of the nation's power grid. Readers will also be offered a bird's eye view on the status of security and system stability and the increasing role of broadband over power lines. In addition, readers will get an overview of environmental standards and issues facing electric utilities and how mergers and acquisitions are affecting the industry as a business enterprise.

For those wanting additional resources, a glossary, acronyms, and an industry contacts directory appear at the end of this book as an appendices.

As will be learned through reading this book, there is much change stirring in the electric utility industry. Many of the brewing changes in the industry are either being postulated or enacted to help strengthen the industry for the future, both in terms of infrastructure for reliability and security sake, as well as in terms of consumer protections for an affordable, stable flow of electricity.

It is my hope that readers can, through the concise and easy-to-read format of this book, learn more about the ever-changing powerful electric utility industry. I hope you enjoy reading through the pages of this book as much as I did creating it for you.

Denise Warkentin-Glenn

Acknowledgments

Thanks to Stephen Hill, Amethyst Hensley, and the rest of the talented team at PennWell for working with me on the manuscript and production processes.

Thanks to my husband, Roger, for all his love, support, prayer, and encouragement. Thanks also to precious Joshua, our toddler son. I love being your mom. Your smile, love, and most of all your presence in my life makes everything more real, more meaningful, and more fun!

Special thanks and praise to God—for without Him I can be and do nothing. (Philippians 4:12-13).

Part I:
The Core Technical Competencies in the Electric Utility Industry

Power Generation

The electric power industry is comprised of many functions. From the generation of electricity flows the transmission and hence the distribution of power to homes, businesses, and other end users. It is, therefore, important to begin at the beginning—generation—particularly because this process happens to be one of the most capital-intensive functions of electric power producers.

To understand the generation of power, it is important to look at the process, the measurement, and the resources required. In general terms, *generation* has a two-fold meaning. First, it may be defined as the process of producing electric energy by transforming other forms of energy. Second, generation can also refer to the amount of electric energy produced for end-user consumption. Expressed in *kilowatt-hours* (kWh), or the electric energy produced in 1 hour (hr) by 1 *kilowatt* (kW) of electric capacity, generation is the amount individual consumers see on their electric bills.

Electricity may be defined as a class of physical phenomena that results from the existence of charge and from the interaction of charges. When a charge is stationary, it produces forces on objects in regions where it is present, and when it is in motion, the charge produces magnetic effects. *Electric and magnetic fields* (EMFs) are caused by the relative position and movement of positively and negatively charged particles of matter. Particles associated with electrical effects are classed as neutral, positive, or negative. However, electricity is concerned with the positively charged particles, such as *protons*, that repel one another. Also of concern are the negatively charged particles, such as *electrons*, which also repel one another. Later in this chapter, the generation process and its magnetic properties, among other aspects of generation, are discussed in detail. However, in summary of this brief discussion, it will suffice to simply state that like charges repel and unlike charges attract.

Electricity is measured in units called *watts*. A watt equals the rate of energy transfer of 1 ampere (A) flowing at a pressure of 1 volt (V) at unity power factor. Electricity is also measured in *watt-hours*, which are equal to 1 watt (W) of power supplied to or taken from an electric current steadily for 1 hr. A kilowatt-hour is equal to 1,000 watt-hours (Wh) used. Homes, businesses, and other end users have their electric power consumption measured by electric, or watt-hour, meters.

In generating electric power, utilities produce power by the megawatt, or 1 million watts. Ratings are used within the industry to denote the amount of electricity each power generating facility is capable of producing. The generation of electricity on a continual basis, at peak performance, is known as the *nameplate capacity* of the generating facility. The continuous, hourly output each generating facility supplies to the electric power grid is termed *net capability*.

Generation Resources

The resources required to generate electricity are immense—particularly in the area of cost to electric utilities. Just as investor-owned utilities (IOUs) spend a lot of money in their operations and maintenance functions, they also happen to capture the lion's share of revenues. In 2003, major IOUs had earned well over $226 million.

As an aside, it is vital to note here, and in later references to financial data, that more recent information was unavailable at the time this book went to press. Data of this nature often lags behind, perhaps in part due to current regulatory schemes governing utility reporting of such data.

As shown in table 1–1, in 2003 electric IOUs spent more than $197 million. In 2000, IOUs spent a whopping $210 million on operations and maintenance and earned more than $235 million.

To understand the variance in expenditures and earnings, one can look at what IOUs spent on purchased power and administrative and general sales for these two years. These are telling dollar figures, in that well over one-half of electric IOU operating revenues is disbursed for expenses, such as the cost of producing electricity and maintaining electric generation plants.

Table 1–2 offers a further look at expenditures. It reveals where major U.S. publicly owned generator and nongenerator electric utilities spent their operations and maintenance dollars.

Table 1–1. Operation and maintenance expenses for major U.S. investor-owned electric utilities

	Million of Nominal Dollars (unless otherwise indicated)		
	1990	1995	2000
Utility Operating Expenses	142,471	165,321	210,324
Electric Utility	127,901	150,599	191,329
Operation	81,086	91,881	132,662
Production	62,501	68,983	107,352
Cost of Fuel	32,635	29,122	32,555
Purchased Power	20,341	29,981	61,969
Other	9,526	9,880	12,828
Transmission	1,130	1,425	2,699
Distribution	2,444	2,561	3,115
Customer Accounts	3,247	3,613	4,246
Customer Service	1,181	1,922	1,839
Sales	212	348	403
Administrative and General	10,371	13,028	13,009
Maintenance	11,779	11,767	12,185
Depreciation	14,889	19,885	22,761
Taxes and Other	20,146	27,065	23,721
Other Utility	14,571	14,722	18,995
Operation (Mills per Kilowatthour)[1]			
Nuclear	10.04	9.43	8.41
Fossil Steam	2.21	2.38	2.31
Hydroelectric & Pumped Storage	3.35	3.69	4.74
Gas Turbine and Small Scale[2]	8.76	3.57	4.57
Maintenance (Mills per Kilowatt hour)[1]			
Nuclear	5.68	5.21	4.93
Fossil Steam	2.97	2.65	2.45
Hydroelectric & Pumped Storage	2.58	2.19	2.99
Gas Turbine and Small Scale[2]	12.23	4.28	3.50

Source: EIA, Electric Power Annual 2003, DOE/EIA-0348(2003) (Washington, D.C., December 2004), Tables 8.1 and 8.2, and EIA, Electric Power Annual 2001, Tables 8.1 and 8.2.

Notes: 1 Operation and maintenance expenses are averages, weighed by net generation
2 Includes gas turbine, internal combusion, photovoltaic, and wind plants

Table 1–2. Operation and maintenance expenses for major U.S. publicly owned generator and nongenerator electric utilities

	Million of Nominal Dollars (unless otherwise indicated)				
	1990	1995	2000	2002	2003
Production Expenses					
Steam Power Generation	3,742	3,895	5,420	6,558	NA
Nuclear Power Generation	1,133	1,277	1,347	1,646	NA
Hydraulic Power Generation	205	261	332	746	NA
Other Power Generation	196	231	603	746	NA
Purchased Power	10,542	11,988	16,481	24,446	NA
Other Production Expenses	155	212	225	1,647	NA
Total Production Expenses[1]	15,973	17,863	24,398	36,188	NA
Operation and Maintenance Expenses					
Transmission Expenses	604	788	982	1,501	977
Distribution Expenses	950	1,274	1,646	1,853	1,044
Customer Accounts Expenses	375	448	662	710	754
Customer Service and Information Expenses	75	120	233	414	311
Sales Expenses	29	29	82	90	95
Administrative and General Expenses	1,619	2,128	2,116	4,058	2,742
Total Electric Operation and Maintenance Expenses	3,653	4,787	5,721	8,625	5,923
Total Production and Operation and Maintenance Expenses	19,626	22,651	30,100	44,813	NA
Fuel Expenses in Operation					
Steam Power Generation	2,395	2,163	4,150	4,818	NA
Nuclear Power Generation	242	222	316	433	NA
Other Power Generation	113	101	373	754	NA
Total Electric Department Employees[2]	N/A	73,172	71,353	93,520	NA

Source: EIA, Financial Statistics of Major US Publicly Owned Electric Utilities 1994, DOE/EIA-0437(94)/2 (Washington, D.C., December 1995), Table 8 and Table 17; EIA, Financial Statistics of Major US Publicly Owned Electric Utilities 1999, DOE/EIA-0437(99)/2 (Washington, D.C., November 2000), Table 10 & Table 21; EIA, Financial Statistics of Major US Publicly Owned Electric Utilities 2000, DOE/EIA-0437(00)/2 (Washington, D.C., November 2001), Table 10 & Table 21; EIA, Public Electric Utility Database (Form EIA-412) 2002; EIA, Electric Power Annual 2003, DOE/EIA-0348(2003) (Washington, D.C., December 2004), Tables 8.3 and 8.4

Notes: 1 Totals may not equal sum of components because of independent rounding.
2 Number of employees were not submitted by some publicly owned electric utilities because the number of electric utility employees could not be separated from the other municipal employees or the electric utility outsourced much of the work.

The number of U.S. electric utilities does fluctuate some with the popular advent of mergers and acquisitions (M&A) within the industry. However, sheer numbers do not necessarily make one type of regulated electric utility more profitable. (More on M&A activity is given in chapter 9.) Table 1–2 clearly shows the relative predominance of IOUs in power generation.

U.S. Department of Energy (DOE) data shows in 2003 total electric utility operating revenues were nearly $314 billion. This represents a 3.2% or $9.8 billion increase as compared to 2002 data. The nation's approximately 240 electric IOUs garnered more than 72% of these revenues.

Table 1–3 shows production, operation, and maintenance expenses for major IOUs and publicly owned electric utilities.

Table 1–4 shows selected electric utility data by ownership, and one may note that there are other utility types represented. Each utility type is discussed in detail in Part IV of this book. For now it will suffice to look at the relative statistical share of sales and revenue that publicly owned, cooperatively owned, and federally owned electric utilities offer to the industry.

Public or municipal electric utilities by and large are distributors of power. However, some of the larger ones produce (generate) and transmit electricity as well. The roughly 2,009 public utilities represent approximately 63% of the number of electric utilities, and supply around 10% of generation and generation capability. This class of utility accounts for 15% of retail sales and around 14% of revenue. Cooperative electric utilities represent approximately 29% of U.S. electric utilities, 9% of sales and revenue, and 4% of generation and generation capability.

Table 1–3. Production, operation, and maintenance expenses for major U.S. investor-owned and publicly-owned utilities (million of nominal dollars)

	Investor-Owned Utilities					Publicly-Owned Utilities[1]				
	1990	1995	2000	2002	2003	1990	1995	2000	2002	2003
Production Expenses										
Cost of Fuel	32,635	29,122	32,555	24,132	26,476	5,276	5,664	7,702	9,696	NA
Purchased Power	20,341	29,981	61,969	58,828	62,173	10,542	11,988	16,481	24,446	NA
Other Production Expenses	9,526	9,880	12,828	7,688	7,532	155	212	225	1,647	NA
Total Production Expenses[2]	62,502	68,983	107,352	90,649	96,181	15,973	17,863	24,398	36,188	NA
Operation and Maintenance Expenses										
Transmission Expenses	1,130	1,425	2,699	3,494	3,585	604	788	982	1,501	977
Distribution Expenses	2,444	2,561	3,115	3,113	3,185	950	1,274	1,646	1,853	1,044
Customer Accounts Expenses	3,247	3,613	4,246	4,165	4,180	375	448	662	710	754
Customer Service and Information Expenses	1,181	1,922	1,839	1,821	1,893	75	120	233	414	311
Sales Expenses	212	348	403	261	234	29	29	82	90	95
Administrative and General Expenses	10,371	13,028	13,009	12,872	13,466	1,619	2,128	2,116	4,058	2,742
Total Electric Operation and Maintenance Expenses	18,585	22,897	25,311	25,726	26,543	3,653	4,787	5,721	8,625	5,923

Source: EIA, Electric Power Annual 2003, DOE/EIA-0348(2003) (Washington, D.C., December 2004), Tables 8.1, 8.3 and 8.4 and EIA, Electric Power Annual 2001, DOE/EIA-0348(2001) (Washington, D.C., December 2002), Table 8.1; EIA, Financial Statistics of Major US Publicly Owned Electric Utilities 1994, DOE/EIA-0437(94)/2 (Washington, D.C., December 1995), Table 8 and Table 17; EIA, Financial Statistics of Major US Publicly Owned Electric Utilities 1999, DOE/EIA-0437(99)/2 (Washington, D.C., November 2000), Table 10 & Table 21; EIA, Financial Statistics of Major US Publicly Owned Electric Utilities 2000, DOE/EIA-0437(00)/2 (Washington, D.C., November 2001), Table 10 & Table 21.; EIA, Public Electric Utility Database (Form EIA-412) 2002.

Notes: 1 Publicly Owned Utilities include generator and nongenerator electric utilities.
2 Totals may not equal sum of components because of independent rounding.

Table 1–4. Selected electric utility data by ownership, 2000

Item	Type of Regulated Electric Utility				
	Investor-owned	Publicly-Owned	Cooperative	Federal	Total[1]
Number of Electric Utilities	240	2,009	894	9	3,152
Electric Utilities (percent)	7.6	63.7	28.4	.3	100.0
Revenues from Sales to Ultimate Consumers (thousand dollars)	169,444,470	33,054,956	20,506,101	1,242,031	224,247,558
Revenues from Sales to Ultimate Consumers (percent)	75.6	14.7	9.1	.6	100.0
Sales of Electricty to Ultimate Consumers (thousand megawatt hours)	2,437,982	516,681	305,856	49,094	3,309,613
Sales of Electricity to Ultimate Consumers (percent)	73.7	15.6	9.2	1.5	100.0
Average Revenue per kWh for Ultimate Consumers (cents)	6.9	6.4	6.7	2.5	6.8
Revenues from Sales for Resale (thousand dollars)	35,359,346	13,430,253	12,027,771	8,900,091	69,717,461
Revenues from Sales for Resale (percent)	50.7	19.3	17.3	12.8	100.0
Sales of Electricity Available for Resale (thousand megawatt hours)	854,228	301,412	311,935	248,664	1,716,239
Sales of Electricty Available for Resale (percent)	49.8	17.6	18.2	14.5	100.0
Average Revenue per kWh for Sales for Resale (cents)	4.1	4.5	3.9	3.6	4.1

1 Includes only those electric utilities in the United States and the District of Columbia.
Note: Totals may not equal sum of components because of independent rounding
Source: Energy Infomration Administration, Form EIA-861, "Annual Electric Utility Report."
Data are based on calendar year submissions.

Generation Efficiency and Capacity

Electric power generators are in a precarious position, due to the somewhat immediate nature of electricity. Because technically electric power cannot be stored, generators must be continually producing power as it is consumed. Consumers of electric power, on the other hand, are obliging in that they produce a relatively constant demand for a basic quantity of electricity at any given time. Referred to as *base load*, all electric utilities have a minimum amount of steadily required electric power that the utility will either purchase or generate itself to meet this demand.

Where do electric utilities obtain their power? If a utility is not in the generation side of the business, most generally electricity is being purchased from a source outside of the utility. Utilities desire to obtain the most efficient and most economical sources of electric power that can be supplied to them. Locale and region play large roles in where and from what sources electric utilities will obtain their electricity. Oftentimes, utilities are purchasing power from more than one source. More on this subject is discussed later in this chapter.

While base load is constant and fairly predictable, electric utilities must be prepared at all times for the sometimes unpredictable demand seasonal weather changes cause. An increase in demand is referred to as *peak load* and *intermediate load*. Typically, peak and intermediate loads are temporary in nature, and utilities generally are flexible enough with an alternate source of power to supply any increases in electric power demand. The vast majority of electric utilities will also have *standby generating capacity*, also known as *capacity margins*, which will be used when demand exceeds supply for a short period of time.

To illustrate how important an electric utility's capacity margins are, one could consider an unusually hot summer or particularly bitter, cold winter. Generation efficiency will drop under these extreme seasonal circumstances. To combat this, utilities will need to have additional sources of electric generation—whether they produce it themselves or purchase it—to meet the temporary increase in electric demand. *Brown outs* can occur should demand for electricity exceed supply, and supply is unable to be generated quickly enough to keep up with the demand.

Should an electric utility have excess capacity, it has the option of selling the "extra" supply of electricity to other utilities in the region through the power grid, otherwise known as *interconnected networks* and the *U.S. bulk power system.*

A complete discussion of the interconnected networks appears in chapter 5. By way of brief introduction, the three interconnected networks are comprised of the Eastern, the Western, and the Texas interconnected systems.

Each electric utility is controlled from a central dispatch center, where adjustments to the production and flow of power are made to match electric utilities' customer needs. When demand exceeds supply, the dispatch center purchases power from the U.S. bulk power system.

The Generation Process

All fuel, regardless of type, possesses potential chemical energy. In turn, this chemical energy is converted into electrical energy. In figure 1–1, the power plant is assumed to be a standard steam power plant. It converts primary fuel into electricity, the boiler turns the water into steam, and the steam turns the fan (turbine) and the generator.

The turbine then rotates the magnet inside the generator. The magnet is surrounded by magnetic lines of force, stretching from one end of the magnet to the other. As the magnet turns, the lines of force are cut by a stationary coil inside the generator, inducing an electric current in the coil.

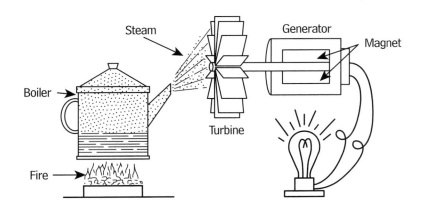

Fig. 1–1. The process of electricity generation

As illustrated in figure 1–1, an electric current is set up in the wire when the magnetic field from the ends of the magnet moves across the turns of wire in the stationary coil. In a standard steam turbine, fuel burns to heat the water until steam is produced. The steam turns a turbine blade, and as the blade of the turbine moves, the magnet inside the generator also turns. If one recalls that the magnet turns within coils of wire, it follows that when the magnet turns, the lines of force it creates cut through the wire and thus induce electric current into the wire. Simply put, this process converts mechanical energy into electrical energy, which is then distributed to the overall electricity grid.

Energy conversion

In order to produce electricity, the generator must have fuel. Generation units, using a variety of technologies, convert energy from falling water, coal, natural gas, oil, and nuclear fuels into electric energy. The majority of electric generators are driven by hydraulic turbines for conversion of fuel energy. And, as discussed earlier, electric power generating plants are interconnected by a transmission and distribution system to serve the electric loads of an area or region. Power transmission and distribution is discussed in chapter 2.

Peak loads, as discussed earlier in this chapter, are generally spikes in usage only lasting a few hours. Peaks are most usually served by gas or oil combustion-turbine or pumped-storage hydropower generating units. The pumped-storage type utilizes the most economical off-peak surplus generating capacity to pump and store water in elevated reservoirs to be released through hydraulic turbine generators during peak load periods. This type of operation improves the capacity factors or relative energy outputs of base-load generating units, hence their economy of operation.

As for the size and capacity of any given electric utility's generating unit or units, there is no *one-size-fits-all*. Both size and capacity of generating units vary widely. They will depend upon the type of unit, duty required (base-, intermediate-, or peak-load service), system size, and the degree of interconnection with neighboring electric systems.

Of extreme importance to electric utilities is the study of annual load graphs and forecasts, which will indicate the rate at which new generation stations must be built to meet demand and expected demand. A look at table 1–5 shows the existing capacity by energy source for data year 2003. Table 1–6 illustrates the planned nameplate capacity additions until the year 2008.

According to the Energy Information Administration's (EIA) long-term forecast, 428 gigawatts (gW) of new generating capacity will be needed by 2025 to meet the growing demand for electricity. This is equivalent to 1,427 new power plants. Adding to this, EIA predicts that most of this new capacity will be fueled by natural gas.

Table 1–5. Existing capacity by energy source (Megawatts)

Energy Source	Number of Generators	Generator Nameplate Capacity (MW)	Net Summer Capacity (MW)	Net Winter Capacity (MW)
Coal[1]	1,535	335,793	313,019	315,237
Petroleum[2]	3,121	40,965	36,429	40,023
Natural Gas	3,069	238,967	208,447	224,366
Dual Fired	3,056	190,739	171,295	183,033
Other Gases[3]	105	2,284	1,994	1,984
Nuclear	104	105,415	99,209	100,893
Hydroelectric[4]	4,145	96,352	99,216	98,399
Other Renewables[5]	1,582	20,474	18,199	18,524
Other[6]	39	704	638	640
Total	16,756	1,031,692	948,446	983,099

1 Anthracite, bituminous coal, subbituminous coal, lignite, waste coal, and synthetic coal.

2 Distillate fuel oil (all diesel and No. 1, No. 2, and No. 4 fuel oils), residual fuel oil (No. 5 and No. 6 fuel oils and bunker C fuel oil), jet fuel, kerosene, petroleum coke (converted to liquid petroleum, see Technical Notes for conversion methodology), and waste oil.

3 Blast furnace gas, propane gas, and other manufactured and waste gases derived from fossil fuels.

4 Conventional hydroelectric and hydroelectric pumped storage. The net summer and winter capacity exceeds the generator nameplate due to upgrades to hydroelectric generators.

5 Wood, black liquor, other wood waste, municipal solid waste, landfill gas, sludge waste, tires, agriculture byproducts, other biomass, geothermal, solar thermal, photovoltaic energy, and wind.

6 Batteries, chemicals, hydrogen, pitch, purchased steam, sulfur, and miscellaneous technologies.

Notes: Where there is more than one energy source associated with a generator, the predominant energy source is reported here. • Totals may not equal sum of components because of independent rounding.

Source: Energy Information Administration, Form EIA-860, "Annual Electric Generator Report."

Table 1–6. Planned nameplate capacity additions from new generators, by energy source, 2004 through 2008 (Megawatts)

Energy Source	2004	2005	2006	2007	2008
Coal[1]	155	991	2,376	4,814	1,390
Petroleum[2]	238	361	344	168	180
Natural Gas	22,490	28,404	23,850	20,985	6,797
Other Gases[3]	–	–	–	580	580
Nuclear	–	–	–	–	–
Hydroelectric[4]	8	11	11	42	4
Other Renewables[5]	257	240	57	36	133
Other[6]	–	–	–	–	–
Total	23,149	30,007	26,638	26,624	9,083

1 Anthracite, bituminous coal, subbituminous coal, lignite, waste coal, and synthetic coal.

2 Distillate fuel oil (all diesel and No. 1, No. 2, and No. 4 fuel oils), residual fuel oil (No. 5 and No. 6 fuel oils and bunker C fuel oil), jet fuel, kerosene, petroleum coke (converted to liquid petroleum, see Technical Notes for conversion methodology), and waste oil.

3 Blast furnace gas, propane gas, and other manufactured and waste gases derived from fossil fuels.

4 Conventional hydroelectric and hydroelectric pumped storage.

5 Wood, black liquor, other wood waste, municipal solid waste, landfill gas, sludge waste, tires, agriculture byproducts, other biomass, geothermal, solar thermal, photovoltaic energy, and wind.

6 Batteries, chemicals, hydrogen, pitch, purchased steam, sulfur, and miscellaneous technologies.

Notes: Where there is more than one energy source associated with a generator, the predominant energy source is reported here. These data reflect plans as of January 1, 2004. Delays and cancellations may have occurred subsequently to the data reporting. • Totals may not equal sum of components because of independent rounding.

Source: Energy Information Administration, Form EIA-860, "Annual Electric Generator Report."

Generation plant circuitry

As shown in figure 1–2, there are a number of main and accessory power circuits in power plants. Main power circuits carry power from generators to the step-up transformers and then to the station high-voltage terminals, whereas auxiliary power circuits provide power to the motors used to drive the necessary auxiliaries.

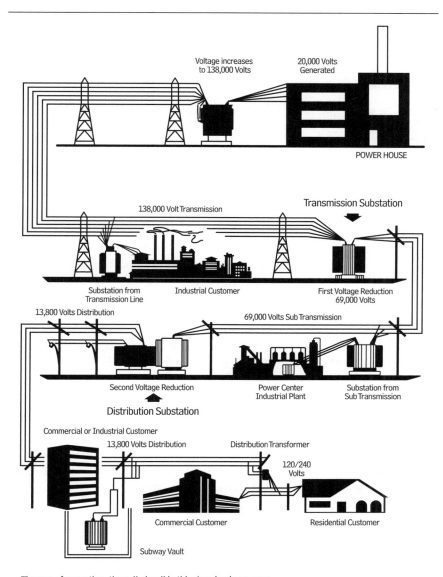

Fig. 1–2. Diagram of the power generation, transmission, and distribution system

There are other circuits needed:

- Control circuits for the circuit breakers and other equipment operated from the plant's control room
- Lighting circuits for the illumination of the plant and to provide power for portable equipment required for power plant maintenance
- Excitation circuits, which must be installed in locations where they will receive good physical and electrical protection, since reliable excitation is required for the operation of the plant
- Instrument and relay circuits for providing values of voltage, current, kilowatts, reactive kilovolt-amperes, temperatures, pressures, flow rates, etc., and to serve the protective system
- Communication circuits, for both plant and system communications (telephone, radio, transmission-line carrier, and microwave radio may be involved)

As can be imagined, the generation plant's power service reliability is at a premium. Because of this fact, the generating station is usually supplied power from two or more sources. To ensure adequate reliability, auxiliary power supplies are often provided for start-up, shutdown, and communications services.

In cases of equipment failure, generation stations must be prepared. Using differential- current and ground relays, over-current relays, and loss-of-excitation relays, generators are immediately (and often automatically) de-energized for electrical failure and shut down for any over-limit condition. Most larger generation plants are monitored constantly by computer- assisted load and frequency control and economic dispatch systems of generation supervision.

Electricity Production Methods

Steam turbines

Steam turbines are the most common method used to produce electricity. A steam turbine plant operation basically consists of four steps. First, water is pumped at high pressure to a boiler. Second, it is heated, most usually by fossil-fuel combustion, to produce steam at high temperature and pressure. Third, this steam flows through a turbine, rotating an electric generator (connected to the turbine shaft), which converts the mechanical energy into electricity.

In the final of the four basic steps, the turbine exhaust steam is condensed by using cooling water from an external source to remove the heat rejected in the condensing process. The condensed water is then pumped back to the boiler to repeat the cycle.

Steam turbine plants can be divided into three categories:

1. Fossil fueled

2. Nuclear

3. Renewable

Fossil-fueled plants are by far the most common. Fossil fuels are of plant or animal origin and consist of hydrogen and carbon (hydrocarbon) compounds. Roughly 70% of the electricity produced in the United States comes from fossil-fueled steam turbine plants. Coal, petroleum, and natural gas are the dominant fossil fuels used in electricity production.

Other fossil fuels utilized include petroleum coke, coke oven gas, and liquefied petroleum. Still yet, there are other types of fossil fuels used for the production of electricity, although they are not commonly used. These other types include peat, oil shale, biomass (wood, etc.), and various waste or by-products, such as steel mill blast furnace gas and refuse-derived fuels.

Of all the fossil fuels, coal is the most widely used. Coal is an inexpensive fuel, as compared to other forms of fuel, and is readily available, since the United States has large deposits. According to the EIA, during 2004, 50% of the nation's electric power was generated at coal-fired plants.

Based upon 2004 net generation shares by energy source, nuclear and natural gas are next in popularity after coal. EIA data shows that there was a 1.8% increase in total net electric power generation for the period between January 1, 2003 and January 1, 2004. This growth in generation was led by natural gas-fired and nuclear generating stations. Natural gas generation increased by 7.6%, which according to EIA reflects the large amount of new gas-fired units electric utilities have installed in recent years. To the surprise of some, nuclear energy is making a comeback. Details of this revelation are included later in this chapter.

Nuclear generation

Nuclear generation increased in 2004 over 2003 usage, as an EIA analysis shows. This increase was primarily due to continued operation of nuclear power plants at high capacity factors and increases in capacity through plant upgrades. Nuclear power accounted for 19.9% of total net generation by the end of 2004.

The resurgence of nuclear power as a generating fuel is significant. As stated above, more details will be provided later in this chapter; however, it is important to have a base knowledge about how nuclear power works.

As presented earlier, figure 1–1 shows the basic process of electricity generation in a typical steam power plant. For purposes of nuclear power generation, however, one should refer to figure 1–3 and figure 1–4. In most electric power plants, water is heated and converted into steam, which drives a turbine-generator to produce electricity. To recap, fossil-fueled power plants produce heat by burning coal, oil, or natural gas.

In a nuclear power plant, the fission of uranium atoms in the reactor provides the heat to produce steam for generating electricity. There are two different reactor processes currently in use: the *pressurized water reactor* (fig. 1–3) and the *boiling water reactor* (fig. 1–4).

Several commercial reactor designs are in use in the United States. The most common is a design that consists of a heavy steel pressure vessel surrounding a *reactor core*. The reactor core contains the uranium fuel. This fuel is formed into cylindrical ceramic pellets about one-half inch in diameter, which are sealed in long metal tubes called *fuel tubes*. The pins are arranged in groups to make a fuel assembly. A group of fuel assemblies forms the core of the reactor.

Heat is produced in a nuclear reactor when neutrons strike uranium atoms, causing them to fission in a continuous chain reaction. Control elements, which are made of materials that absorb neutrons, are placed among the fuel assemblies. When the control elements, or control rods as they are often called, are pulled out of the core, more neutrons are available, and the chain reaction speeds up, producing more heat. Whey they are inserted into the core, more neutrons are absorbed, and the chain reaction slows or stops, reducing the heat.

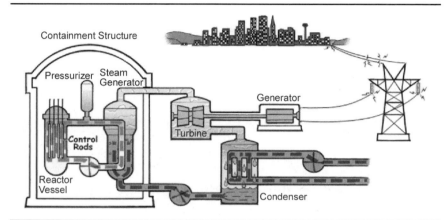

Fig. 1–3. Pressurized water reactor

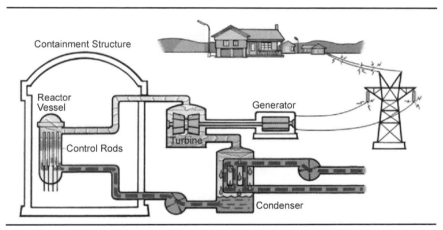

Fig. 1–4. Boiling water reactor

Most commercial reactors use ordinary water to remove the heat created by the fission process. These are called *light water reactors*. The water also serves to slow down, or moderate, the neutrons. In this type of reactor, the chain reaction will not occur without the water to serve as a moderator. Figures 1–3 and 1–4 show the two different types of light water reactor designs currently in use.

In a pressurized water reactor (PWR), the heat is removed from the reactor by water flowing in a closed pressurized loop. The heat is transferred to a second water loop through a heat exchanger. The second loop is kept at a lower pressure, allowing the water to boil and create steam. This steam is used to turn the turbine-generator and produce electricity. Afterward, the steam is condensed into water and returned to the heat exchanger.

In a boiling water reactor (BWR), water boils inside the reactor itself, and the steam goes directly to the turbine-generator to produce electricity. As in a PWR, the steam is condensed and reused.

Gas turbine and combined-cycle generating plants

Power plants with gas turbine–driven electric generators are often used to meet short-term peaks in electrical demand. Gas turbine power plants use atmospheric air as the working medium, operating on an open cycle where air is taken from and discharged to the atmosphere and is not recycled.

As shown in figure 1–5, in a simple gas turbine plant, compressed gas is ignited, and the hot gases rotate a gas turbine, generating electricity. Put another way, air is compressed and fuel is injected into the compressed air and burned in a combustion chamber. Variations of this basic operation to increase cycle efficiency include regeneration, where exhaust from the turbine is used to preheat the compressed air before it enters the combustion chamber.

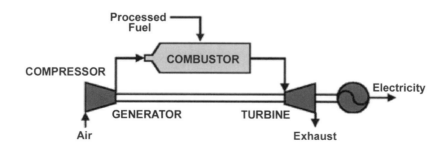

Fig. 1–5. Simple cycle gas turbine

The principle behind how gas turbines work is similar to a jet engine. Gas turbine generation is often used for peak, emergency, and reserve power production because of their quick start-up time. The downside is that gas turbines tend to be less efficient than their steam turbine counterparts. Generally, gas turbines are 100 MW or less. However, some are more, and they may be installed in a wide range of locations.

Gas turbines generally require smaller capital investments than coal or nuclear and can be designed to generate small or large amounts of power. The main advantage of gas turbines is the ability to be turned on and off within minutes, supplying power during peak demand. Large turbines may produce hundreds of megawatts.

Gas turbines may be described thermodynamically by the Brayton cycle (see fig. 1–6). Air is compressed isentropically, combustion occurs at constant pressure, and expansion over the turbine occurs isentropically back to the starting pressure.

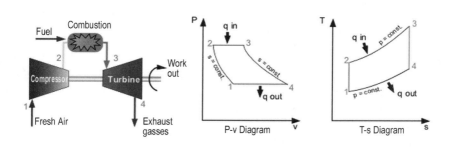

Fig. 1–6. Brayton cycle

As with all cyclic heat engines, higher combustion temperature means greater efficiency. The limiting factor is the ability of the steel, ceramic, or other materials that make up the engine to withstand heat and pressure. Considerable engineering goes into keeping the turbine parts cool.

Most turbines also try to recover exhaust heat to the compressed air, prior to combustion. Combined-cycle designs pass waste heat to steam turbine systems. Combined heat and power or cogeneration uses waste heat for hot water production. Both combined cycle and cogeneration are discussed later in this chapter.

Mechanically, gas turbines can be less complex than internal combustion piston engines. Simple turbines might have one moving part—the shaft/compressor/turbine/alternator-rotor assembly, not counting the fuel system. More sophisticated turbines may have multiple shafts (spools), hundreds of turbine blades, movable stator blades, and a vast system of complex plumbing, combustors, and heat exchangers.

The largest gas turbines operate at 3,000 or 3,600 rpm to match the AC power grid. They require a dedicated building. Smaller turbines, with fewer compressor/turbine stages, spin faster. Jet engines operate around 10,000 rpm, and microturbines around 100,000 rpm.

Thrust and journal bearings are a critical part of design. Traditionally, they have been hydrodynamic oil bearings, or oil-cooled ball bearings. This is giving way to hydrodynamic foil bearings, which have become commonplace in microturbines (discussed later) and auxiliary power units (APUs).

Power plant gas turbines can range in size from truck-mounted mobile plants to enormous, complex systems. For example, one could consider the GE H Series power generation gas turbine. This 400-MW unit has a rated thermal efficiency of 60% when waste heat from the gas turbine is recovered by a conventional steam turbine in a combined cycle.

According to the U.S. Department of Energy (DOE), with the restructuring of the industry, increasing numbers of power companies are planning units in the 30- to 200-MW range. The DOE forecasts that about one-half the U.S. demand for gas turbine systems through 2020 is likely to be for midsize turbines, suitable for both central and distributed power applications. Besides providing both steady electricity and meeting surges in power demand, these smaller turbines might also be ideal for repowering aging coal plants and relieving congestion in the power transmission system.

In order to make gas turbines more efficient, many electric utilities utilizing gas turbines for peak, emergency, or reserve power production will have hybrid generation units. These specialized units combine the benefits of gas and steam turbines. Known as *combined-cycle generating units*, the efficiency at which they can operate is far better than just using the gas turbine technology alone. Figure 1–7 is a simplified schematic of a combined-cycle generating unit.

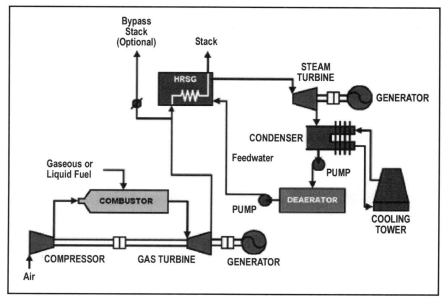

Fig. 1–7. Combined-cycle generating unit diagram

In a combined-cycle power plant, a gas turbine generator is combined with a steam turbine power plant with the overall objective of increasing the efficiency of electricity generation. In a thermal power plant, high-temperature heat as input to the power plant is converted to electricity as one of the outputs and low-temperature heat as another output. In order to achieve high efficiency, the temperature of the input heat should be as high as possible. And, as a rule, the temperature of the output heat should be as low as possible.

For gas turbine generators, the input temperature to the gas turbine is approximately 900°C to 1,200°C. This is a relatively high temperature; however, the output temperature of the flue gas is also rather high, at some 450°C to 650°C. For steam turbine power plants, the output temperature of the cooling water is significantly lower (20°C to 40°C), but the input temperature to the steam turbine is also significantly lower (420°C to 580°C).

Therefore, by combining both processes through the combined-cycle generation process, high input temperatures and low output temperatures can be achieved, and power plant efficiency can be increased.

The output heat of the gas turbine flue gas is utilized to generate steam in a heat recovery steam generator (HRSG) and therefore is used as input heat to the steam turbine power plant. The HRSG can be designed with or without supplementary firing. Without supplementary firing, the efficiency of the combined-cycle power plant is higher. With supplementary firing, the plant is more flexible to respond to fluctuations of electrical load.

Whereas gas turbines usually are fired by relatively expensive fuels, such as natural gas, gas from coal gasification, or light fuel oil, the HRSG can also be fired by less expensive fuels, such as heavy fuel oil or coal.

In the case of generating only electricity, power plant efficiencies of up to 58% can be achieved. In the case of combined heat and power generation, efficiency increases to about 5%.

Cogeneration. Also known as combined heat and power or CHP, this process uses a power station to simultaneously generate both heat and electricity. CHP allows a more total use of energy than conventional generation, potentially reaching an efficiency of 70% to 90%, as compared with around 50% for conventional plants. This means less fuel needs to be consumed to produce the same amount of energy.

Thermal power plants (including those using uranium, coal, petroleum, or natural gas) do not convert all available energy into electricity. Inevitably, a large amount of heat is released as a by-product. Conventional power stations emit this heat into the environment through cooling towers, as flue gas, or by other means.

To utilize this waste heat in places where it would otherwise need to be generated by other means is the most energy efficient usage. Often, these other means involve drawing upon electric power, while it would be more efficient to supply the heat directly, without first converting it into electricity. Heat is widely used, not only for residential buildings, but also for high-temperature industrial processes and other applications.

Cogeneration systems are generally economic on a large scale, for instance to provide heating water and power for an industrial site or an entire town. There are several common CHP plant types:

- Gas turbine CHP plants using the waste heat in the flue gas of gas turbines
- Combined-cycle power plants adapted for CHP
- Steam turbine CHP plants using the waste heat in the steam after the steam turbine

Small cogeneration units for hospitals, swimming pools, or groups of dwellings are also economic if standardized, mass-produced CHP plants are used. Examples are the internal combustion (IC) engines (gas or diesel engines) used for car manufacture. They use the waste heat in the flue gas and cooling water of a gas or diesel engine and replace the traditional gas- or oil-fired boiler (furnace) used in central heating systems.

Generating high-temperature heat (i.e., from industrial processes) usually results in some wasted low-temperature heat. This is usually emitted into the environment. A CHP system can also be used to recover some of this waste heat and to use it to generate electric power. For small systems, the waste heat can be recovered. One challenge with cogeneration is heat transmission over long distances. Thick, heavily insulated pipes are required, whereas electricity can be transmitted along a comparatively simple wire.

Microturbines. Microturbine technology is becoming widespread for distributed power and generator applications. Distributed generation is a new trend in electric power generation. The concept permits electric consumers who generate their own electricity to send their surplus electrical power back into the power grid. Microturbines range from handheld units producing less than 1 kW to power station units producing megawatts.

To understand why this technology has been such a hit for distributed power producers, it is important to dig a little deeper into their popularity. Part of the success of microturbines is due to advances in electronics, which allow unattended operation and interfacing with the commercial power grid. Electronic power switching technology eliminates the need for the generator to be synchronized with the power grid. This allows, for example, the generator to be integrated with the turbine shaft, and to double as the starter motor.

Another advantage of microturbines is that the technology's alternators have high power density in relation to volume and width, as compared with piston engines. This is due in large part to high rotation speed. The need for a recuperator, however, does mitigate this advantage.

Microturbine designs usually consist of a single-stage radial compressor, a single-stage radial turbine, and a recuperator. Recuperators are difficult to design and manufacture because they operate under high pressure and temperature differentials. Waste heat can be used for hot water production. Typical microturbine efficiencies are 20% to 35%. When in a combined heat and power system, overall efficiencies of greater than 90% can be achieved.

For the aforementioned reasons, there is little wonder why the technology has become so popular, particularly for distributed generators. The reason for the existence of distributed generation follows.

Most usually distributed generation is utilized by factories, offices, and especially hospitals, which require extremely reliable sources of electricity and heating for air conditioning and hot water. To safeguard their electricity supply and to reduce their costs, some institutions, factories, and offices install cogeneration or total energy plants, often using waste material, such as wood waste or surplus heat from an industrial process, to generate electricity.

In some cases, electricity is generated from a locally supplied fuel, such as natural gas or diesel oil, and the waste heat from the generator's thermal energy source is then used to provide hot water and industrial heating as well. It is often economic to have a cogeneration plant when an industrial process requires large amounts of heat that are generated from nonelectric sources, such as fossil fuels or biomass (these and other fuels are discussed later in this chapter).

Until recently, regulatory and technology issues meant domestic consumer-generated electricity could not be easily or safely coupled with the incoming electric power supply. Electric utilities need to have the ability to isolate parts of the power grid; when a line goes down, workmen have to be sure the power is off before they work on it. Utilities also spend much effort maintaining the quality of power in their grid. Distributed power installations can make control of these issues more difficult.

With the advent of extremely reliable power electronics, it is becoming economic and safe to install even domestic-scale cogeneration equipment. These installations can produce domestic hot water, home heating, and electricity. Surplus energy can then be sold back to the power company. Advances in electronics have eased electric companies' safety and quality concerns. Regulators can act to remove barriers to the uptake of increased levels of distributed generation by ensuring that centralized and distributed generation are operating on a level playing field.

To be sure, distributed generation is not limited to fossil fuels. Some countries and regions already have significant renewable power sources in power grid–tied wind turbines and biomass combustion, both of which are examined later in this chapter.

Increasing amounts of distributed generation will require changes in the technology required to manage transmission and distribution of electricity. There will be an increasing need for network operators to manage networks actively rather than passively. Increased active management will bring added benefits for consumers in terms of the introduction of greater choice with regard to energy supply services and greater competition. However, the switch to a more active management, currently under way, will likely be a difficult one. Distribution networks are considered natural monopolies, and are thus tightly regulated to ensure they do not draw excess profits at the expense of consumers. As will be discussed later in this book, network investment will be a key determinant of the costs networks can pass on to consumers.

As an aside, networks act to maximize their profits within the framework provided by their regulation. Currently such regulation does not lend itself very well to offering incentives for innovative behavior by networks, although some of that behavior is cropping up. This likely will prove to be a barrier both to the development of the networks and to increases in the levels of distributed generation that are added to networks.

On the upside, there are indications of regulatory authorities becoming more aware of the potential barriers. Regulation of connection charges and conditions has been introduced to enable distributed generators to participate in the electricity market.

There is the potential for a major portion of the electricity power supply to come from decentralized power sources. Thus billing and energy credits, generation control, and system stability remain significant issues limiting the widespread use of this technology. To maintain control and stability of the power system in some networks, neighboring consumers need to consume all the electric power a producing consumer may produce. This ensures there is a net flow of electric power from generators to consumers in the distribution network, even though there may be a local outflow within the local distribution.

With the continued growth of electricity markets and the requirement for open access to networks, the distributed generator may have more options for selling the excess production, either through physical or financial contracts (hedges).

Generation and the Role of Fuel Diversity

Fuel diversity refers to the variety and proportions of energy sources used to produce electric power. By having a variety of energy sources available, risk and opportunity can be spread across a variety of fuels, taking advantage of emerging technologies and resources. At the same time, it provides a buffer for electric companies and consumers from fuel price swings, fuel unavailability, and changes in regulatory policies.

According to the Edison Electric Institute (EEI), the trade group representing electric IOUs, a diverse mix of fuels is used across the county to generate electricity. These regional differences are shown in figure 1–8. Electric utilities choose their fuel mixes, according to EEI, using price and supply availability as the primary determinants.

EEI statistics show electricity consumption will increase 54% by 2025. Diverse fuel supply options will thus be key to keeping the power flowing in the most reliable, cost-effective, and efficient ways to use and conserve energy.

Nature's power

From the foregoing discussions, it is clear that coal, natural gas, oil, and now even nuclear power all have roles to play in the U.S. energy-producing economy. However, there are pronounced concerns regarding how some of these fuel types affect the environment. Thus it has become the goal of not only the U.S. president, but others as well, that the United States must take full advantage of both renewable and alternative energy sources.

According to the May 2001 Report of the National Energy Policy Development Group (NEPDG), "A sound national energy policy should encourage a clean and diverse portfolio of domestic energy supplies. Such diversity helps to ensure that future generations of Americans will have access to the energy they need."

The NEPDG was charged with developing a wide range of recommendations for the future of energy and energy production. More on the recommendations and the National Energy Policy can be found in chapter 6.

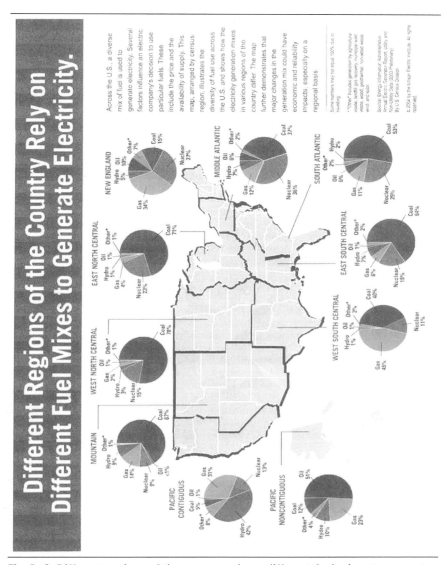

Fig. 1–8. Different regions of the country rely on different fuel mixes to generate electricity

Renewable energy is generated from five sources:

- Biomass
- Hydropower
- Geothermal
- Wind
- Solar

While the current contribution of renewable and alternative energy resources to the total U.S. electricity supply is only around 9%, renewable and alternative energy sectors are among the fastest growing in the United States. This is not to say that their contribution will surpass that of coal, oil, nuclear, or natural gas as fuel sources, but only that renewable and alternative energy have a significant role in powering the United States.

Alternative energy includes alternative fuels that are transportation fuels other than gasoline and diesel, even when the type of energy, such as natural gas, is traditional. It also includes the use of traditional energy sources, such as natural gas, in untraditional ways, such as for distributed energy at the point of use through microturbines or fuel cells. Finally, it also encompasses future energy sources, such as hydrogen and fusion.

Renewable energy taps naturally occurring flows of energy to produce electricity, fuel, heat, or a combination of these energy types. Hydropower has long provided a significant contribution to the U.S. energy supply and is competitive with other forms of conventional electricity.

Nonhydropower energy is generated from biomass, geothermal, wind, and solar. The United States has significant potential for renewable resource development. These renewable sources of energy are domestically abundant and often have less impact on the environment than such fossil fuels as coal and natural gas.

Renewable energy technologies

Renewable and alternative energy sources can be produced centrally or on a distributed basis, near their point of use. Providing electricity, light, heat, or mechanical energy at the point of use reduces the need for some transmission lines and pipelines, reducing associated energy delivery losses and increasing energy efficiency. Next are some descriptions of renewable energy technologies.

Biomass power. Also called biopower, biomass power is electricity produced from biomass fuels. Biomass consists of plant materials and animal products. There are many sources of biomass fuels:

- Residues from the wood and paper products industries
- Residues from food production and processing
- Trees and grasses grown specifically as energy crops
- Gaseous fuels produced from solid biomass, animal wastes, and landfills

Unlike other renewable energy sources, biomass can be converted directly into liquid fuels, called *biofuels*, to meet transportation needs. The two most common biofuels are ethanol and biodiesel. Ethanol is made by fermenting any biomass rich in carbohydrates, such as corn. It is mostly used as a fuel additive to reduce vehicle emissions. Biodiesel is made using vegetable oils, animal fats, algae, or even recycled cooking greases. It can be used as a diesel additive to reduce emissions or in its pure form to fuel vehicles. Beyond energy benefits, development of biomass benefits rural economies that produce crops used for biomass, particularly ethanol and biomass electricity generation.

Biopower technologies convert renewable biomass fuels into electricity (and heat) using modern boilers, gasifiers, turbines, generators, and fuel cells. Biomass can be converted into electricity in one of several processes. However, most biomass electricity is generated using a steam cycle, as shown in figure 1–9. In the direct combustion/steam turbine system, biomass is burned in a boiler to make steam. Then the steam turns a turbine, which is connected to a generator that produces electricity.

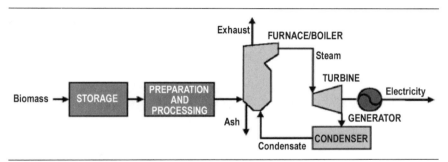

Fig. 1–9. Direct combustion/steam turbine system

In a conventional power plant, biomass can also be burned with coal in high-efficiency boilers to produce steam and electricity. Cofiring biomass with coal is an affordable way for electric utilities to obtain some of the environmental benefits of using renewable energy.

Solid biomass can be converted into a fuel gas using a biomass gasification system. In this method, sand (at about 1,500°F) surrounds the biomass and creates a very hot, oxygen-starved environment. These conditions break apart wood or other biomass and create an energy-rich, flammable gas. The biogas can be cofired with wood (or other fuel) in a steam boiler or used to operate a standard gas turbine.

Biogas can also be created by digesting food or animal wastes in the absence of oxygen. Called anaerobic digestion, this process will occur in any airtight container containing a mixture of bacteria normally present in animal waste. Different types of bacteria work in sequence to break down complex chemicals, such as fat and protein, into progressively simpler molecules. The final product is a biogas containing methane and carbon dioxide. The biogas can be used for heating or for electricity generation in a modified internal combustion engine. However, it should be noted that advanced gasification technologies are necessary for converting animal waste to a biogas with sufficient energy to fuel a gas turbine.

Landfills also produce a methane-rich biogas from the decay of wastes containing biomass. However, landfill gas must be cleaned to remove harmful and corrosive chemicals before it can be used to generate electricity.

This "clean" landfill gas, or methane, can be captured and burned in a boiler to produce steam for electricity generation or for industrial processes.

It is interesting to note that fuel gases made from biomass can be used to generate electricity in a gas turbine (see fig. 1–5) or in a combined-cycle generating unit (see fig. 1–7). As discussed earlier, in a simple-cycle gas turbine, compressed gas is ignited, and the hot gases rotate a gas turbine, generating electricity. In a combined-cycle unit, the hot waste gases from the gas turbine are used to create steam to run a steam turbine and generator.

U.S. biopower plants have a combined capacity of 7,000 MW. These plants use around 60 million tons of biomass fuels from primarily wood and agricultural wastes to generate approximately 37 billion kWh of electricity each year. Biomass is becoming rather popular with electric utilities. It accounts for around 76% of renewable electricity generation and about 1.6% of total U.S. electricity supply.

As with conventional power from fossil fuels, biopower is available continuously. Small, modular biopower systems with rated capacities of 5 MW or less can supply power in regions without grid electricity. These systems can provide distributed power generation in areas with locally produced biomass resources, such as rice husks or walnut shells.

According to the DOE, biomass provides new markets for the nation's farmers and creates jobs in rural communities. At the same time, biomass is good for the balance of trade between the United States and other countries. The United States is developing small, clean, and efficient biomass power plants for villages and small communities in developing countries as well as for the developed world. This expertise in small biopower systems abroad is paving the way for an expanded role in distributed generation at home.

DOE also maintains that biopower is a natural fit for the electric power industry. Power plants co-firing biomass with coal have fewer acid-rain producing emissions. In addition, biomass has some characteristics in common with fossil fuels, such as being available on demand.

In spite of its relative advantages, biopower does have some disadvantages. Biopower plants have higher generation costs than those of fossil fuel generation. Biomass fuels contain less concentrated energy, are less economic to transport over long distances, and require more preparation and handling than fossil fuels. Challenges facing the increased use of biopower include competition with natural gas, the need to develop high-yield, low-input energy crop farming practices, and the need for more research to improve biopower technologies.

Geothermal power. Geothermal energy technologies utilize the Earth's heat for direct-use applications, geothermal heat pumps, and electrical power production. Research in all areas of geothermal development is helping to lower costs and expand its use. Most geothermal resources in the United States are found in the West; however, it should be noted that geothermal heat pumps can be used anywhere.

Because geothermal energy uses heat from the Earth, it is clean and sustainable. Resources range from the shallow ground to hot water and hot rock found miles beneath the Earth's surface. Geothermal resources can also be found deeper in the Earth's surface to the extreme high temperatures of molten rock, called magma.

Shallow ground, or the upper 10 feet of the Earth's surface, maintains a temperature of between 50°F and 60°F. Geothermal heat pumps tap into this resource to heat and cool buildings.

The heat pump system consists of a heat pump, an air delivery system (ductwork), and a heat exchanger. The heat exchanger is nothing more than a system of pipes buried in the shallow ground near the building. In the winter, the heat pump removes heat from the heat exchanger and pumps it into the indoor air delivery system. In the summer, the process is reversed, and the heat pump moves heat from the indoor air into the heat exchanger. The heat removed from the indoor air during the summer can also be used to provide a source of hot water.

For the generation of electricity, wells are drilled into underground reservoirs. Some geothermal power plants use steam from a reservoir to power a turbine/generator, while others use the hot water to boil a working fluid that vaporizes and then turns a turbine. Hot water near the surface of the Earth can be used directly for heat. Direct-use applications include heating buildings, growing plants in greenhouses, drying crops, heating water at fish farms, and several other industrial processes, such as pasteurizing milk.

Hot, dry rock resources occur at depths of 3 to 5 miles everywhere beneath the Earth's surface, and at lesser depths in certain areas. Access to these resources involves the process of injecting cold water down one well, circulating it through hot fractured rock, and drawing off the heated water from another well. As of this writing of this book, there are not any commercial applications of this technology. Additionally, existing technology allows for the recovery of heat directly from magma, the very deep and most powerful resource of geothermal energy.

Improvements in drill bits, drilling techniques, advanced instruments, and other technological advances have made energy production from geothermal resources increasingly efficient. Geothermal resources account for approximately 17% of renewable electricity generation and around 0.3% of total U.S. electricity supply. The utilization of geothermal resources as an energy source is significant, given the fact that the net installed capacity of geothermal plants has increased from 500 MW in 1973 to 2,800 MW today.

Hydropower. Hydropower is another source of electricity that also uses turbines. In the case of hydro, the turbine is turned by the kinetic energy of water.

Hydroelectric power generation represented 6.6% of U.S. electric power in 2004. Conventional hydro generation, the largest renewable generation source, actually declined in 2004 by 2.2 % over 2003 usage because of drought conditions, most notably in the West.

There are three basic types of hydro: falling water, run of river, and pumped storage systems. The traditional and most common hydro system consists of a dam with a reservoir behind it. This is the *falling water* type, where water falls through conduits, referred to as penstocks, and turns turbines connected to generators.

The *run of the river* type is based upon the force of a river's current turning turbines and has no reservoir. This makes run of the river generating plants dependent upon seasonal changes in river flow. Pumped storage takes advantage of off-peak periods to pump water up into a reservoir with electric pumps. When additional power is needed, the water can be released to turn turbines and produce power.

These types of power generation are often used to meet base load demand. Steam turbine plants are most efficient where they are operated on a continual basis. This is at least due in part to the fact that these plants do not produce power until the water is hot enough to boil. For immediately available peak load, utilities use gas turbine units, internal combustion engines, and hydroelectric units, which can respond quickly to changes in demand.

Gas turbines operate by passing the hot gasses produced from combustion of natural gas or oil directly through a turbine. These units are generally 100 MW or less (some are more, as discussed above) and are less efficient than steam turbine units. Internal combustion engines, such as diesel generators, are portable and instantaneous sources of electricity used for emergency situations as well as for reserve electricity. These may be 5 MW or less in size; however, there are some internal combustion engines that may be larger.

All of these methods of electric generation—steam turbines, gas combustion turbines, water turbines, wind turbines, and internal combustion engines—are referred to as *prime movers*. Some generating units can use more than one type of fuel. These units are known as *dual-fired units* and may be either sequentially fired or concurrently fired. *Sequential plants* use one fuel, then the other, while *concurrent plants* can use two fuels, such as coal and natural gas, at the same time.

Ocean energy. Drawing upon the energy of ocean waves, tides, or on the thermal energy (heat) stored in the ocean, ocean energy is abundant. The ocean contains two types of energy— *thermal energy* from solar heat and *mechanical energy* from the tides and waves. It has been estimated that if 0.2% of the ocean's untapped energy could be harnessed, it could provide enough power for the entire world.

Oceans cover more than 70% of the Earth's surface, making them the world's largest solar collectors. The sun warms the surface water a lot more than the deep ocean water, and this temperature difference stores thermal energy. Thermal energy is used for many applications, including electricity generation.

Currently, three types of electricity conversion systems exist. They are *closed-cycle, open-cycle,* and *hybrid.* Closed-cycle systems use the ocean's warm surface water to vaporize a working fluid, which has a low boiling point, such as ammonia. The vapor expands and turns a turbine. The turbine then activates a generator to produce electricity. Open-cycle systems actually boil the seawater by operating at low pressures. This produces steam that passes through a turbine/generator. And finally, hybrid systems combine both closed-cycle and open-cycle systems.

Ocean mechanical energy is different from ocean thermal energy. Even though the sun affects all ocean activity, tides are driven primarily by the gravitational pull of the moon, and waves are driven primarily by the winds. A barrage, or dam, is used to convert tidal energy into electricity by forcing the water through turbines, activating a generator.

For wave energy conversion, there are three basic systems:

- Channel systems that funnel the waves into reservoirs
- Float systems that drive hydraulic pumps
- Oscillating water column systems that use the waves to compress air within a container.

The mechanical power created from these systems either directly activates a generator or transfers to a working fluid, water or air, which then drives a turbine/generator.

Solar power. Sunlight or solar energy can be used to generate electricity, heat water, and heat, cool, and light buildings. Photovoltaic (solar cell) systems use semiconductor materials similar to those used in computer chips to capture the energy in sunlight and to convert it directly into electricity. Photovoltaic (PV) cells have been used in everything from the solar cells in calculators to Space Station Freedom.

In a PV cell, the solar energy knocks electrons loose from their atoms, allowing the electrons to flow through the material to produce electricity. Typically, PV cells are combined into modules that hold about 40 cells. About 10 of these modules are mounted in PV arrays, which can be used to generate electricity for a single building, or in the case of a large number of PV arrays, for a power plant.

A power plant can also use a *concentrating solar power system*, which uses solar heat to generate electricity. The sunlight is collected and focused with mirrors to create a high- intensity heat source. This heat source produces steam or mechanical power to run a generator that creates electricity.

Solar water heating systems for buildings have two main parts—a solar collector and a storage tank. A *flat-plate collector*, a thin, flat, rectangular box with a transparent cover, is mounted on the roof, facing the Sun. The Sun heats an absorber plate in the collector, which in turn heats the fluid running through tubes within the collector. To move the heated fluid between the collector and the storage tank, a system either uses a pump or gravity, as water has a tendency to naturally circulate as it is heated. Systems using fluids other than water in the collector tubes usually heat the fluid by passing it through a coil of tubing in the tank.

Many large commercial buildings can use solar collectors to provide more than just hot water. Solar process heating systems can be used to heat these buildings, and a ventilation system can be used in cold climates to preheat air as it enters a building. Heat from a solar collector can even be used to provide energy for cooling a building.

A solar collector is not always needed when using sunlight to heat a building. Some buildings can be designed for *passive solar heating*. These buildings usually have large, south-facing windows. Materials that absorb and store solar heat can be built into the sunlit floors and walls. The floors and walls will then heat up during the day and slowly release heat at night —a process called *direct gain*. Many of the passive solar heating design features also provide *daylighting*. This is the use of natural sunlight to brighten up a building's interior.

Some architects are using careful design and new optical materials to use sunlight to reduce the need for traditional lighting and to cut down on heating and cooling costs. For example, materials that absorb and store solar heat can be built into the sunlit floors and walls. The floors and walls will then store heat during the day and slowly release it at night.

Another technology for harnessing solar energy is a concentrating solar power system. In this system, the Sun's heat is used to generate electricity. Sunlight is collected and focused with mirrors to create a high-intensity heat source, which in turn can be used to generate electricity through a steam turbine or a heat engine.

Solar energy accounts for roughly 1% of renewable electricity generation and around 0.02% of total U.S. electricity supply. While solar energy technologies have undergone technological and cost improvements through the years, continued research is needed to reduce costs and improve performance. It is important to note, however, that solar technologies are well established in high-value markets, such as remote power, satellites, communications, and navigational aids.

Wind energy. Wind energy has been used since at least 200 B.C. for grinding grain and pumping water. By 1900, windmills were used on farms and ranches in the United States to pump water and, eventually, to produce electricity. Windmills have developed into what is currently termed *wind turbines*.

Wind turbines are used for several applications. Wind power uses naturally occurring wind energy for practical purposes, like generating electricity, charging batteries, or pumping water. Electric utilities can purchase wind power, often from outside sources, such as wind energy farms. Wind farms produce electricity using large modern wind turbines operating together. Homeowners, farmers, and remote villages can use small wind turbines to help meet localized energy needs.

Wind turbines capture energy by using propeller-like blades mounted on a rotor. These blades are placed on top of high towers. This is done in order to take advantage of the stronger winds at 100 feet or more above the ground. The wind causes the propellers to turn, which then turn the attached shaft to generate electricity.

Wind can be used as a stand-alone energy source or in conjunction with other renewable energy systems. Wind and natural gas hybrid systems are a promising approach, offering consumers a clean alternative to conventional energy generating systems.

Wind energy accounts for approximately 6% of renewable electricity generation and 0.1% of total electricity supply. However, with the help of research labs, universities, electric utilities, and wind energy developers, advancements in wind power technology have helped to curb wind energy's costs by 85% during the last 20 years.

In certain parts of the United States (i.e., the West, the Great Plains, and New England), wind power can be produced at prices comparable to other conventional energy technologies, making wind energy particularly ripe for considerable growth in the future. Additionally, incentives like the federal production tax credit and net metering provisions available in some areas are improving the overall economics of wind energy.

New Technologies

The continuing presence of federal regulations has placed more and more strongholds on the electric utility industry with respect to emissions controls, safety issues, and cost containment issues (to name a few). Thus the need for newer, faster, less expensive, and more environmentally friendly technologies has, and will continue, to surface.

Electric utilities face a plethora of existing as well as new regulations, requiring—even demanding—technology to step up to the plate and address the overriding issues of clean, safe, reliable, and affordable electricity production.

Nuclear revival?

It appears as though there is a considerable amount of renewed interest in nuclear-generated power. In short, this has occurred because nuclear power is comparatively inexpensive and fits the bill as a source of "clean" energy. And certain new technologies, which will be discussed below, have made a nuclear comeback even more attractive to some.

To be certain, nuclear in general has never totally lost its appeal to many electric utilities, most of which was cast toward the relatively few emissions issues, but nuclear has had a tough go in the electric utility industry.

After the near-disaster at Three Mile Island in 1979 and the 1986 catastrophe at Chernobyl, support for nuclear power all but vanished entirely. Most wrote it off as too dangerous to have any real place in the current U.S. energy economy. For the most part, nuclear power plants were once thought to be too expensive to build, too complex to operate, and their radioactive emissions too dangerous and costly.

Now, three decades later, the industry is seeing an about-face on the issue of nuclear power. Electric utilities are considering building or restarting up to eight reactors in Mississippi, South Carolina, Alabama, Virginia, Idaho, and Illinois. Recent polls show that 50% of Americans now support the construction of new nuclear power plants. This figure includes 59% of energy-strapped Californians.

Support for nuclear is strong on Capitol Hill. The Energy Policy Act of 2005 revives nuclear power as one of the up-and-coming fuel sources of the future. For starters, the act proposes to offer subsidies to utilities constructing new nuclear power plants. More on the energy plan can be found in chapter 6.

Chicago-based Exelon leads the way in the return of nuclear to the electric power industry. The company happens to be the nation's largest nuclear operator, with 17 in operation in Illinois, Pennsylvania, and New Jersey. Why is Exelon pushing nuclear? Well, the company claims that nuclear is a much-needed fuel source for power generation. However, it is also because there has been a technological shift to a new generation of cheap, efficient reactors. Exelon claims it would produce lots of power without generating the greenhouse gases that fossil-fuel plants emit.

Exelon has approached the Nuclear Regulatory Commission (NRC) about building a new generation of safer, smaller mini-reactors. The NRC has not had a nuclear reactor construction license to consider since 1978. Before forging ahead with any construction plans, however, Exelon (as well as any other companies wanting to build new plants) will need to seek NRC's approval. And, should they gain approval, rules and regulations on the construction, operation, and maintenance of the new reactors will be ultra-stringent. This will be due in part because of the disasters and near-disasters for which nuclear is best known. Also, the rules and regulations will be strictly laid down and enforced to help allay the general public's fears about the use of nuclear power, containment, and waste disposal issues.

At the forefront of the technology Exelon proposes to employ is the *pebble bed modular reactor* or PBMR. Proponents of PBMR call it "remarkable." Opponents of the technology call it "unproven" and merely a remake of an unsuccessful base technology that has met an untimely death in several countries, including England, France, and Germany, as well as in the United States.

The PBMR's basic design concept revolves around the high-temperature gas-cooled reactor (HTGR). The design concept has been hailed for the last 30 years as inherently safe. However, there have been problems with the HTGR design. For example, there were the 1967 and 1989 closures of the Peach Bottom Unit 1 and Fort St. Vrain reactors in the United States.

The most recent PBMR project is a hybrid of past efforts at utilizing the HTGR design concept. The effort is piloted by an international conglomerate of U.S.-based Exelon Corporation (Commonwealth Edison, PECO Energy, and British Energy), British Nuclear Fuels Limited, and South African-based ESKOM as merchant nuclear power plants. The consortium, called Nustart, has preliminary plans to search for two new sites to build plants utilizing the HTGR hybrid design. Nustart proposes to seek licensing for a new plant in 2008, with construction beginning in 2010.

Unlike light water reactors that use water and steam, the PBMR design would use pressurized helium heated in the reactor core to drive a series of turbine compressors attached to an electrical generator. The helium is cycled to a recuperator to be cooled down and returned to cool the reactor, while the waste heat is discharged into the environment. Designers of the technology claim there are no accident scenarios that would result in significant fuel damage and catastrophic release of radioactivity.

The safety claims are grounded on the heat-resistant quality and integrity of some 400,000 tennis-ball sized graphite fuel assemblies, or pebbles, which are continuously fed from a fuel silo through the reactor bit by bit to keep the reactor core only marginally critical. Each spherical fuel element has an inner graphite core embedded with thousands of smaller fuel particles of enriched uranium (up to 10%) encapsulated in multiple layers of nonporous, hardened carbon. The fuel's slow circulation through the reactor provides for a small core size. This minimizes excess core reactivity and lowers power density.

Opponents claim the lack of a containment building is one of the biggest drawbacks of the PBMR design. Not having a containment building does save money; however, some in the industry contend this makes the technology unsafe. Proponents claim the containment building is not in the PBMR design plan because it would hinder the design's passive cooling feature of the reactor core through natural convection.

Other possible design features, which may increase health and safety risks, include the lack of an emergency core cooling system, and a reduced one-half mile emergency planning zone. In comparison, the light water reactor design has a 10-mile emergency planning zone. The Nuclear Information and Resource Service estimates that a single 110-MW PBMR would produce 2.5 million irradiated fuel elements during a 40-year operational cycle. Naysayers of the PBMR technology say this fact, as well as other issues, would make radioactive contamination uncertainties surrounding the technology persist long after a PBMR has closed.

Clean coal technologies

Clean coal technologies (CCT) are the products of research and development conducted over the past 20 years. The results are more than 20 new, lower-cost, more efficient, and environmentally compatible technologies for electric utilities, steel mills, cement plants, and other industries. Several of these technologies are detailed a little later in this chapter.

According to the Coalition for Affordable and Reliable Energy (CARE), CCT helped make it possible for U.S. utilities to meet more stringent Clean Air Act (CAA) requirements, while continuing to utilize America's most plentiful domestic energy resource.

Originally, the CCT program focused on commercializing processes to help reduce sulfur dioxide (SO_2) and nitrogen oxide (NO_x) emissions. Begun in 1986, the original program was aimed at demonstrating more efficient and environmentally friendly alternatives to traditional pulverized coal boilers.

Newer programs in CCT are essential for building on the progress of the original CCT program, according to CARE. New programs are also vital in the process of finding solutions for reducing trace emissions of mercury, reducing or eliminating carbon dioxide (CO_2) emissions, and increasing fuel efficiencies. One such program is the Clean Coal Power Initiative (CCPI), an industry/government partnership to implement the president's National Energy Policy recommendations to increase CCT investment.

Over the longer term, CCT technology research will be directed toward developing coal-based hydrogen fuels. CARE predicts these coal-based hydrogen fuels coupled with sequestration will allow for greater use of coal with zero emissions. To this end, the DOE has announced a presidential initiative to build FutureGen, a $1 billion project that will lead to the world's first emission-free plant to produce electricity and hydrogen from coal while capturing greenhouse gases.

Coal is keeping pace with consumer electric demand. According to EIA, electricity demand will increase some 53.4% over the next 25 years. Meeting this rising growth rate will require the construction of the equivalent of more than 1,200 new power plants of 300 MW each. Many of these new plants will be coal-fired. This equates to around 65 power plants being built each year.

Coal has been keeping pace with electric demand, and has for the most part grown its generation supply base since at least the early 1990s. EIA's long-term energy outlook suggests coal will account for 51% of power generation in 2025. These new technologies hold promise to help meet this increased demand, and to continue the decline in SO_2 and NO_x emissions already taking place.

The CCT innovations contributing to making environmental protection possible fall into three categories:

1. Combustion
2. Postcombustion
3. Conversion

In combustion, coal is combined with other substances in the boiler to improve efficiency and remove any impurities. For instance, in fluidized-bed combustion, limestone or dolomite are added during the combustion process to reduce SO_2 formation.

Postcombustion refers to using *scrubbers* (flue gas desulfurization), chemical cleaning, or precipitators to remove large quantities of sulfur, other impurities, and particulate matter (dust and ash) from emissions before they are released into the atmosphere.

Conversion is the process of using heat and pressure to convert coal into a gas or liquid, which can be further refined and used cleanly. An example would be *integrated gasification combined cycle*, and other gasification and liquefaction technologies.

A little history on the CCT program shows programs were selected and technology demonstrations conducted between 1986 and 1993. These demonstrations were jointly funded efforts by the government and industry to demonstrate and commercialize new, lower cost options for controlling SO_2 and NO_x emissions at coal-fired power plants. Efforts were also aimed at improving power plant performance and efficiencies.

The program proved successful. It resulted in 35 pioneering projects in 17 states, which eventually produced 22 commercial successes. Commercial success, according to DOE, occurs with those technologies and programs resulting in domestic or international sale of the technology, or the technology continues to operate economically at the plant site. To see a detailed list by project, company, and location, one can access the following Web site: www. netl.doe.gov/cctc/.

Some of the major technological benefits of CCT by technology are presented next.

Low nitrogen-oxide burners. These specialized NO_x burners are currently installed on around 75% of U.S. coal-based power plants. They helped electric utilities to comply with more stringent CAA NO_x emission reduction requirements. In addition, between 1980 and 2000, NO_x emissions from coal power plants declined 56%, measured by pounds of emissions per kWh. Low NO_x burners have played an increasing role in these lowered emissions.

Selective catalytic reduction (SCR). This technology achieves NO_x reductions of 80% to 90%, and costs roughly one-half of what it did in the 1980s. SCR systems are on order or are under construction on 30% of the existing U.S. coal-fired generating capacity.

Flue gas desulfurization (FGD). More than 400 commercial FGD units have been deployed at one-third of their 1970s cost. SO_2 emissions from coal-based power plants declined 61% between 1980 and 2000, based on pounds of emissions per kWh. This was despite the fact that utilities increased their use of coal by 74% over the same time period. FGD systems were credited with having played a significant role in making this happen.

Fluidized bed combustion (FBC). There are more than 170 operating FBC units in the United States and 400 worldwide. The technology, which removes SO_2 and NO_x inside the boiler with no additional controls required, has been commercialized. More than $6 billion in domestic sales and nearly

$3 billion in overseas sales have resulted from U.S. public and private investment in FBC technology, research, development, and demonstration.

Integrated gasification combined cycle (IGCC). Currently, there are more than 1,500 MW of coal-based generation operating today. Another 1,900 MW are gasifying refinery wastes, and an additional 2,200 MW are in the design state.

Fuel cells

Fuel cells use the chemical energy of hydrogen to generate electricity. When pure hydrogen is used, the only by-products are pure water and useful heat. Fuel cells are unique in terms of the variety of their potential applications. They can provide energy for systems as large as a utility power station, as small as a laptop computer, and just about everything in between, including light cars and trucks.

A fuel cell consists of an electrolyte and two catalyst-coated electrodes (a porous anode and cathode). Several types of fuel cells are currently under development. Each type of fuel cell is classified according to the kind of electrolyte it uses. This is because the electrolyte ultimately determines which applications the fuel cell is most suitable for, as well as its advantages and limitations. Different electrolytes determine the kind of chemical reactions taking place in the cell and the temperature range in which the cell operates.

The DOE hydrogen program, which is focused on transportation, supports research and development of *polymer electrolyte membrane* (PEM) fuel cells. PEM fuel cells currently are primarily used for light-duty vehicles. Research to advance other types of fuel cells, including molten carbonate and solid oxide fuel cells, is also underway within DOE.

In 2003, the president of the United States announced a $1.2 billion hydrogen initiative to reverse America's growing dependence on foreign oil and to reduce greenhouse gas emissions. The president urged the development of commercially viable hydrogen fuels and technologies for cars, trucks, homes, and businesses.

Since that time, a variety of research and development issues have emerged. One such instance was in 2004, when DOE announced a new phase of fuel cell research designed to hasten the wider availability of zero-emissions energy. With this announcement, DOE presented 11 new projects with a price tag of nearly $4.2 million that will focus on solving the remaining issues in developing solid oxide fuel cell systems for commercial use.

The U.S. Secretary of Energy said in a report on new fuel cell projects, "The President's Hydrogen and Climate Initiatives envision fuel cells playing a prominent role in the economy and everyday life. To reach the goal of zero-emissions energy, we need to reduce the costs of fuel cell acquisition and use. These projects address the last barriers to commercially viable solid oxide fuel cell systems."

Anatomy of a fuel cell. As shown in figure 1–10, which is a basic fuel cell diagram, a fuel cell power system has many components. The heart of the fuel cell is the *fuel cell stack*. In the basic operation, fuel enters a reformer, creating a process fuel, which typically is hydrogen or a hydrogen-rich fuel. The fuel then passes to the anode. Depending on the type of fuel cell, positive or negative ions are exchanged with the cathode, yielding free electrons and an exhaust product.

Next, electrons from the anode cannot pass through the membrane to the cathode, so they travel around it through a DC electric current. Air is supplied to the cathode to complete the reaction. Water vapor and heat are generated as a by-product of the reaction. In some cases, heat recovery systems can use the heat for other processes. Finally, the power conditioner converts DC power to AC power for consumer use.

According to the DOE, fuel cells have the potential to replace the internal combustion engine in vehicles and provide power for stationary and portable power applications. They can be used in transportation applications, such as powering automobiles, buses, cycles, and other vehicles. Many portable devices can be powered by fuel cells, such as laptop computers and cell phones. They can also be used for stationary applications, such as providing electricity to power homes and businesses.

The benefits of fuel cells are several fold. They are cleaner and more efficient than traditional combustion-based engines and power plants. When pure hydrogen is used to power a fuel cell, the only by-products are water and heat. Because fuel cell technology is more efficient than combustion-based technologies, less energy is needed to provide the same amount of power.

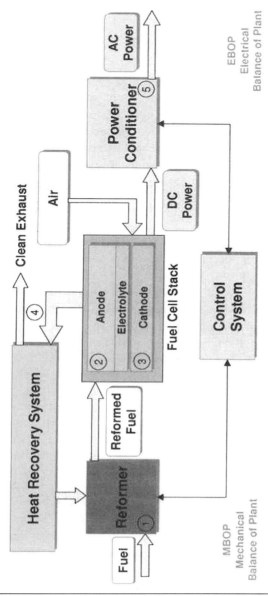

Fuel Cell Stack

AC Power

Power Conditioner ⑤

DC Power

Clean Exhaust

Air

Heat Recovery System

Reformed Fuel

Reformer ①

Fuel

Anode ②
Electrolyte
Cathode ③

Control System

MBOP
Mechanical
Balance of Plant

EBOP
Electrical
Balance of Plant

1) Fuel enters a reformer that creates a process fuel, typically hydrogen or a hydrogen rich fuel.

2) Hydrogen rich fuel passes to the anode. Depending on the type of fuel cell, positive or negative ions are exchanged with the Cathode yielding free electrons and an exhaust product.

3) Electrons from the anode cannot pass through the membrane to the cathode, so they travel around it

through a DC electric circuit. Air is supplied to the cathode to complete the reaction.

4) Water vapor and heat are generated as a by-product of the reaction. In some cases, Heat Recovery Systems can use the heat for other processes.

5) The Power Conditioner converts DC Power to AC Power for use by the customer.

© 2003 Caterpillar Inc.

Fig. 1–10. Basic fuel cell diagram

Also, hydrogen can be produced using a wide variety of resources found in the United States (including natural gas, biological material, and even water). Consequently, using hydrogen fuel cells reduces U.S. dependence on other countries for fuel.

In addition to the technical challenges being addressed through research, design, and development, DOE contends there are obstacles to successful implementation of fuel cells and the corresponding hydrogen infrastructure that can only be addressed by integrating the components into complete systems. After a technology achieves its technical targets in the laboratory, the next step is to show it can work as designed within complete systems— for example, fuel cell vehicles and hydrogen refueling infrastructure.

DOE maintains that technology validation has confirmed that component technologies can be incorporated into a complete system solution, and system performance and operation are met under anticipated operating scenarios. In addition to developing and testing complete system solutions for transportation and infrastructure, DOE is addressing the technology for the production of electricity for consumers. Development and testing is ongoing in what DOE terms real-world operating conditions and realistic operating conditions. There are several challenges to overcome in paving the way for commercialization of fuel cell and hydrogen infrastructure technologies. These include fuel cell cost and durability, hydrogen storage, hydrogen production and delivery, and public acceptance.

The role of hydrogen. Hydrogen does not exist alone in nature. Natural gas contains hydrogen (about 95% of natural gas is methane), as does biomass and hydrocarbons, such as coal. An equally diverse array of primary energy sources, such as wind, solar, geothermal, nuclear, and hydropower, can be used to extract hydrogen from water.

Primary energy sources are found or stored in nature. They include biomass, coal, oil, natural gas, sunlight, wind, water, nuclear power, geothermal power, and potential energy from the Earth's gravity. Energy carriers are not energy sources. They are produced from primary energy sources using technology. These include the electricity produced from coal or photovoltaics, and ethanol produced from corn. In this last example, the resource (corn) from which the energy carrier (ethanol) is extracted is called a *feedstock*. Hydrogen is an energy carrier that can be produced from a wide variety of feedstocks. This diversity of options enables hydrogen production almost anywhere in the world.

All hydrogen production processes are based on the separation of hydrogen from hydrogen-containing feedstocks. The feedstock dictates the selection of the separation method. Currently, two primary methods are used to separate hydrogen—thermal and chemical. A third method, biological, is in the exploratory research and development stage.

Currently, 95% of the hydrogen produced in the United States, around 9 million tons per year, uses a thermal process with natural gas as the feedstock. This process, called steam methane reformation (SMR), consists of two steps. First, the feedstock undergoes reformation with high-temperature steam supplied by burning natural gas to obtain a synthesis gas. Second, a water-gas shift reaction to form hydrogen and carbon dioxide from the carbon monoxide produced in the first step is used.

To a lesser degree, the United States also produces hydrogen electro-chemically from water when higher purity hydrogen is needed. The process, called electrolysis, passes electricity through water in an ionic transfer device to separate water into its hydrogen and oxygen parts. Renewable energy technologies, such as wind turbines, can generate electricity to produce hydrogen from electrolysis with zero greenhouse gas emissions. In France, an abundance of nuclear power makes electrolysis a logical (and the most common) method for producing hydrogen.

Hydrogen production infrastructure. The development of a national hydrogen production infrastructure to support a *hydrogen economy* could evolve along one or more pathways, according to DOE. For example, this could include a distributed production infrastructure located at the point of use, or a centralized production infrastructure at large industrial production sites.

While *distributed hydrogen production* requires smaller capital investments and a minimal transport and delivery infrastructure, centralized production achieves the economic benefits of mass production. *Power parks*, which produce electricity and hydrogen, typically produce hydrogen during off-peak hours so they can provide electricity during high grid loads or blackouts. Power parks are another production pathway for providing transportation fuel.

The DOE's Hydrogen, Fuel Cells & Infrastructure Technologies Program responds to several recommendations in the president's National Energy Policy. There are several features included in this program:

- The development of next-generation technologies
- Establishment of an education campaign touting potential benefits
- Improved integration of subprograms in hydrogen, fuel cells, and distributed energy

Guided by the National Hydrogen Energy vision, the Hydrogen, Fuel Cells, & Infrastructure Technologies Program works in partnership with industry, academia, and national laboratories, as well as other DOE programs. The goal is to overcome technical barriers through research and development of hydrogen production, delivery, and storage technologies, as well as fuel cell technologies for transportation, distributed stationary power, and portable power applications. The program also addresses safety concerns and develops model codes and standards. It aims to validate and demonstrate hydrogen and fuel cell technologies in real-world applications, and to educate key stakeholders whose acceptance of these technologies will determine their success in the marketplace.

A hydrogen economy. A new hydrogen economy will require cost-effective hydrogen production and expanded hydrogen infrastructure to ensure convenient access to hydrogen energy to end users. Like many other technologies, research, development, and demonstration must continue to lower cost, increase efficiency, and address emissions issues associated with some hydrogen production technology.

The transition to a hydrogen economy features a variety of processes from a diverse resource base. Currently, according to DOE, the U.S. transition will likely build on the existing infrastructure and begin with a mix dominated by fossil fuels in the near term. This will be followed by an increased presence of renewable energy resources and possibly nuclear energy.

In conclusion, energy systems of the mid- to late-21st century will need to be cleaner and much more efficient, flexible, and reliable than they are currently. This will ensure U.S. energy security and environmental viability. Hydrogen and fuel cell technologies have the potential to solve the major energy security and environmental challenges that the United States faces—dependence on petroleum imports, poor air quality, and greenhouse gas emissions. Overcoming the technical barriers existent with these new technologies is a paramount concern of the U.S. government. It is no surprise that much of DOE's current research and development efforts and dollars are being set aside for just this purpose.

Renewables

Renewable energy technologies have enormous potential in the United States. This, according to the Union of Concerned Scientists (UCS), is punctuated with what UCS and others in the industry say are significant market barriers, which may limit renewable development in the future unless special policy measures are enacted.

In particular, wind energy technology has been leading the way in recent years among the varied sorts of renewable energy. Twenty-five years ago, wind turbine capacities were at a mere 50 kW. Currently, that number is up to 2,000 kW. The average size of a utility-scale wind turbine is 1.2 MW.

Wind energy has also managed in the current times to reduce the cost of energy by more than 80%. A single 1.5-MW wind turbine can power hundreds of homes. Adding to this, wind farms are being built more and more around the world. Total wind energy capacity at the end of 2004 was 47,317 MW. The United States generated 6,740 MW of this total capacity. That was enough power to serve 1.6 million average-size homes.

The DOE's wind program, based at the National Wind Technology Center (NWTC) in Golden, Colorado, is working to make wind energy increasingly popular as a viable source of power. According to NWTC, in order to become competitive, wind turbines will need to be even larger and more efficient than the machines in existence today. The bulk of the some 6,740 MW of wind capacity is generated by technologies requiring higher wind speeds—6.7 meters per second (m/s) or 15 mph at a height of 10 meters (m) or 33 feet (ft).

The development of land-based wind energy technologies would be ideal, since these technologies can produce electricity in low wind speed areas. This development could be key to ensuring the industry's growth over the next three to six years. NWTC's goal is to help the industry develop these land-based technologies to produce electricity in lower wind speed areas. This goal is ambitious, as is the 3¢/kWh target by 2012.

However, NWTC maintains that bigger and more efficient wind turbine machines are only part of what can make wind energy a more viable power source. For example, there would be enough wind in the Great Plains area to generate more electricity than the United States currently uses. The barrier in this scenario would be that the U.S. transmission grid is already heavily loaded and could not support the addition of that much wind energy.

According to NWTC, long-term industry growth will depend upon the development of deepwater, offshore wind technologies. It has numerous advantages:

- Higher quality wind resources with reduced turbulence and increased wind speed
- Load proximity
- Increased transmission options
- Potential for reducing land use and aesthetic concerns
- Relaxed size constraints on transportation and installation

These advantages give offshore technologies an edge over the traditional land-based ones. NWTC offshore objectives are to focus on technologies needed for deepwater wind energy applications, such as cost-effective, highly reliable floating platforms and anchoring systems.

Renewable electricity standards. In an effort to help reduce market barriers and stimulate new markets for all types of renewable energy sources, a growing number of states have adopted *renewable electricity standards* (RES). An RES requires electric utilities to gradually increase the amount of renewable energy resources—such as wind, solar, and bioenergy—in their electricity supplies.

At the time of writing this book, 18 states and Washington, D.C. have implemented minimum RES requirements. In 2004, Colorado voters passed the first-ever RES ballot initiative requiring the state's utilities to generate 10% of their electricity from renewable energy sources by 2015.

In September 2004, New York created the second-largest new renewable energy market in the country, behind only California, when the state's Public Service Commission adopted an RES of 24% by 2013. Hawaii, Maryland, Pennsylvania, Rhode Island, and Washington, D.C. also enacted minimum RES standards in 2004. In addition, 8 states enacted an RES as part of electricity generation deregulation legislation, and 10 states enacted standards outside of utility restructuring. Several states, including Minnesota, Nevada, New Mexico, New Jersey, and Pennsylvania, have revisited and significantly increased or accelerated their standards.

According to UCS projections, state RES laws and regulations will provide support for more than 25,550 MW of new renewable power by 2017—an increase of 192% over total 1997 levels (excluding hydro). This represents enough clean power to meet the electricity needs of 16.9 million typical homes. The standards in California, New York, Pennsylvania, and Texas create the four largest markets for renewable energy growth, according to UCS. By 2017, new annual renewable energy production from all state RES programs will reduce CO_2 emissions by 64.3 million metric tons. This level of reduction is equivalent to taking 9.6 million cars off of the road or planting more than 15.4 million acres of trees.

At this point, it is difficult to know the true effectiveness of RES programs. However, a number of studies have found renewable energy resources standards to be the primary driver of new renewable energy generation in the United States. Two-thirds of the wind development installed between 1998 and 2003 (3,300 MW) occurred in states having an RES. In Minnesota, Xcel Energy has acquired about 600 MW of wind and bioenergy as a result of its requirement. Wisconsin utilities have secured enough renewable resources to meet their state's target through 2011, and Iowa has met and exceeded its relatively low RES requirement.

The Texas Legislature in 1999 adopted an RES requiring 2,000 MW of new renewable electricity generating capacity to be installed by 2009. The RES was signed into law by the state's governor and was implemented by the Federal Energy Regulatory Commission (FERC). Reportedly, more than 1,100 MW of renewable energy have already been installed in Texas, which puts the state well ahead of its 2005 target of 850 MW.

Why has Texas been so successful in its efforts? UCS suggests the state's success is due in part to the availability of good renewable energy resources within the state. The inclusion of several key provisions in the legislation has also helped bring about success. These provisions include new renewable energy requirements high enough to trigger market growth in the state, and requirements can be met using tradable renewable energy credits. Also, requirements apply across the board to all electricity providers, and retail providers not complying with the RES target must pay significant financial penalties.

In states where utilities divested generation and credit-worthy power marketers have not emerged (as in the Northeast), or utilities have had credit problems (as in Nevada), new renewable energy projects have had a difficult time obtaining contracts and financing. Many of these states are addressing the issues by creating new supplemental mechanisms, such as using state agencies to provide financing or credit price guarantees.

From the foregoing discussion, it is clear states have demonstrated RES effectiveness. In addition, several U.S. surveys show Americans strongly favor clean, renewable energy sources and support a national RES standard. Because investments in renewable energy create important benefits for the entire nation, it follows that the RES could become a cornerstone of U.S. national energy policy.

Power Transmission and Distribution

The second process of delivering electricity, behind the generation of power by power plants, is transmission. However, as discussed in detail below, it is the electric energy transmission and distribution (T&D) networks working together that ultimately deliver electricity to consumers.

While power generation plants produce electric power, and bulk power transmission systems route the electric power to distribution systems, the distribution systems round out the process to deliver electricity to retail customers. As noted in chapter 1, the generation component is the most expensive piece of net electric utility plant investments, at around 55%. Transmission represents 12%, and distribution roughly 29%.

While electric power generation takes up the largest portion of electric utility investment capital, all components are essential. The bulk power transmission system enables electric utilities to deliver power over long distances. This capability increases the potential for competition by providing electricity customers an opportunity to purchase less expensive power from distant suppliers. Figure 2–1 gives a basic look at how electric power transmission operates from the power plant to *substations*, which will be discussed below, and ultimately to consumers via power lines.

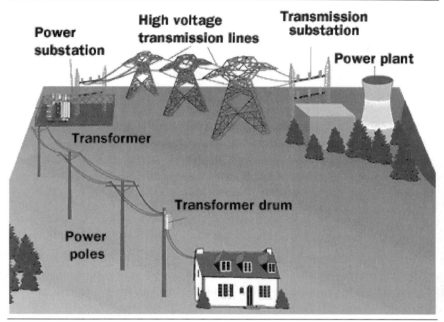

Fig. 2–1. Basic electric power transmission flow

The Bulk Power Transmission System

The U.S. bulk transmission grid, which is divided into three main power grids, allows generation facilities to produce large quantities of energy and then deliver it to distribution networks for delivery to retail customers for consumption. As shown in figure 2–2, the three main power grids are the Eastern Interconnect, the Western Interconnect, and the Texas Interconnect.

As depicted in figure 2–2, there are 10 North American Reliability Council (NERC) regions:

- ECAR—East Central Area Reliability Coordination Agreement
- ERCOT—Electric Reliability Council of Texas
- FRCC—Florida Reliability Coordinating Council
- MAAC—Mid-Atlantic Area Council

- MAIN—Mid-America Interconnected Network
- MAPP—Mid-Continent Area Power Pool
- NPCC—Northeast Power Coordinating Council
- SERC—Southeastern Electric Reliability Council
- SPP—Southwest Power Pool
- WSCC—Western Systems Coordinating Council

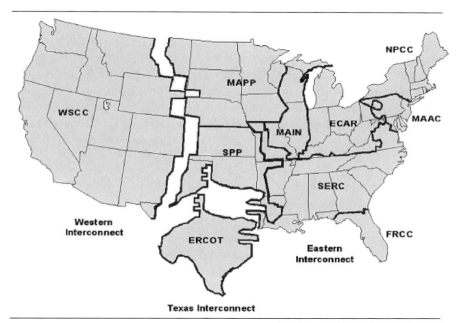

Fig. 2–2. U.S. power grids

By way of brief explanation, the Eastern and Western interconnects have limited interconnections with each other, and the Texas Interconnect is only linked with the others through DC lines. Both the Western and the Texas interconnects are linked with Mexico, and the Eastern and Western interconnects are strongly interconnected with Canada. All electric utilities in the mainland United States are connected to at least one other utility via these power grids.

The grid systems of Hawaii, Alaska, and Puerto Rico are of special note. They are much different than those on the U.S. mainland. Alaska has an interconnected grid system, but it connects only to Anchorage, Fairbanks, and the Kenai Peninsula. Much of the rest of the state depends upon small diesel generators, although there are a few minigrids in the state as well. Hawaii also depends upon minigrids to serve each island's inhabitants. Minigrids, which are a distributed energy application, will be examined later in this chapter. Finally, Puerto Rico has a network of more than 500 MW of installed capacity, which connects only to island inhabitants.

Further, the interconnections are divided into 152 regional control areas for the express purpose of the safe, dependable, and reliable operation of the three main electric network systems. The main function of the control areas is to monitor and control a regional transmission grid. Control areas also are the primary units responsible for the reliable operation of the transmission system. More details on the regulatory oversight of bulk power transmission owners and the development of regional transmission organizations (RTOs) and independent system operators (ISOs) is discussed in chapter 4 and chapter 5, respectively.

The transmission system is the central trunk of the electricity grid. Thousands of distribution systems branch off from this central trunk and diverge into tens of thousands of feeder lines. These feeder lines provide electric power to homes, buildings, and industries. The power flow to the distribution systems (discussed later in this chapter) is largely determined by the power flow through the transmission systems.

To be sure, the transmission system is truly a grid. Transmission lines run not only from power plants to load centers, but also from transmission line to transmission line, providing a redundant system to help ensure smooth power flow. This redundancy is imperative for a number of reasons. One of the most important ones is that if a transmission line is taken out of service in one part of the grid, the power can usually then be rerouted through other power lines to continue delivering uninterrupted power to consumers.

Put another way, the power from many power plants is pooled in the transmission system, and each distribution system draws from this pool of power. This networked system helps achieve a high reliability for power delivery because any one power plant needing to shut down will only constitute a fraction of the power being delivered by the grid.

One of the results of power pooling is electricity drawn off the grid always comes from a diversity of power sources. These may include coal, nuclear, natural gas, oil, and renewable energy sources (see chapter 1). This is often referred to as *system power,* since it is the standard power mixture supplying the transmission system.

Power grid components and input

A transmission grid is comprised of power stations, transmission circuits, and substations. Energy is usually transmitted with three-phase AC. The voltage level on the bulk power transmission sytem is typically between 115 kV and 765 kV.

At the generating plants, electric power is produced at a generally low voltage of up to 25 kV, then stepped up by the power station transformer to a higher voltage for transmission over long distances to grid exit points (substations). It is necessary to transmit the electricity at high voltage to reduce the percentage of energy lost. For a given amount of power transmitted, a higher voltage reduces the current and resistance losses in the conductor. Long-distance transmission is typically at voltages of 100 kV and higher. Transmission voltages up to 765 kV AC and up to +/– 533 kV DC are currently used in long-distance overhead transmission lines.

In an AC transmission line, the inductance and capacitance of the line conductors can be significant. The currents flowing in these components of the transmission line impedance result in the generation of *reactive power.* Reactive power does not transmit any energy to the load. Reactive current flow causes extra losses in the transmission circuit.

The fraction of total energy flow (power) that is resistive (as opposed to reactive) power is known as the *power factor.* Electric utilities will often add capacitor banks and other measures throughout the system to control reactive power flow for reduction of losses and stabilization of system voltage. These measures may include phase-shifting transformers, static volt ampere reactive (VAR) compensators, and flexible AC transmission systems.

High-voltage direct current (HVDC) is used to transmit large amounts of power over long distances or for interconnections between asynchronous grids. When electrical energy is required to be transmitted over very long distances, it can be more economical to transmit using DC rather than AC.

For a long transmission line, the value of the smaller losses and the reduced construction cost of a DC line can offset the additional cost of converter stations at each end of the line. Also, at high AC voltages, significant amounts of energy are lost due to *corona discharge*. This is the capacitance between phases, or in the case of buried cables, between phases and the soil or water in which the cable is buried. Since the power flow through an HVDC link is directly controllable, HVDC links are sometimes used within a grid to stabilize the grid against control problems with the AC energy flow.

At the substations, transformers are again used to step the voltage down to a lower voltage for distribution to commercial and residential users. This distribution is accomplished with a combination of subtransmission (34.5 kV to 115 kV) and distribution (4.6 kV to 25 kV). Finally, at the point of use, the energy is transformed to low voltage (100 V to 600 V).

Transmission lines can also be used to carry data. This is referred to as power-line carrier (PLC), also called broadband over power lines (BPL), and is discussed in chapter 7. Briefly, PLC signals can easily be received with a radio for the longwave range. Sometimes there are also communications cables using the transmission line structures. These are generally fiber optic cables, and often they are integrated into the underground conductor. Sometimes a stand-alone cable is used, which is commonly fixed to the upper crossbar on the power pole.

Power quality and reliability

Today's power grid faces more challenges than ever before. Chapter 7 addresses this topic more fully, but it is important to note here some of the more common power quality issues affecting the bulk power grid. Additional information is given in the latter part of this chapter concerning some of the DOE's suggested solutions to the national power crisis. The crisis is the overloaded power grid. There is much more demand on the bulk power grid in the current times than ever before. DOE's proposed solutions generally revolve around new technologies to address the important issues facing the U.S. power grid.

Through the growth of the manufacturing industry and the increased use of electricity by other business sectors and household consumers, a number of power quality issues have surfaced. Due to the sensitive nature of computers and computer-operated pieces of equipment, particularly in the business sectors, the most severe power quality problem is a large voltage surge caused by a lightning strike.

There may be other causes of power quality problems:

- **Voltage sags and swells.** Large electrical motors switching on and off can cause widely varying sags and surges in electrical AC load. Both sags and surges are the most commonly experienced power quality problem among electronic and computer equipment users.

- **Impulse events.** Generally referred to as glitches, spikes, or transients, these events may occur repeatedly and may or may not follow a pattern.

- **Decaying oscillatory voltages.** The voltage deviation gradually dampens, like a ringing bell. This is caused by banks of capacitors being switched on by an electric utility.

- **Commutation notches.** These are caused by momentary short circuits.

- **Harmonic voltages.** These can be present at very high frequencies to the extent they cause equipment to overheat and interfere with the performance of the sensitive electronic equipment.

Other power quality problems can be considered liability problems because they occur when the transmission system is not capable of meeting the load on the system. These may include several problems:

- **Brownouts.** These are persistent lowerings of system voltage caused by too many electrical loads on the transmission line.

- **Blackouts.** These are a complete loss of power. Most usually, surprise blackouts are caused by equipment failures, such as a downed power line, a blown transformer, or a failure in the relay circuit. Although normally limited by design to small geographic areas, blackouts have been know to affect wide regions of the United States.

- **Rolling blackouts.** These are intentionally imposed upon a transmission grid when electrical load exceeds generation capabilities. By blacking out a small sector of the grid for a limited amount of time, some of the load on the grid is removed, allowing the grid to continue serving the rest of the customers. To spread the burden among customers, the sector that is blacked out is changed every 15 minutes or so, and thus the blackouts roll through the service area.

Power quality is clearly a major issue electric power utilities face when dealing with their customers. Thus the next several pages of this book will be devoted to types of transmission lines, distribution system facilities, and the link that ties them all together.

Transmission and Distribution Components

Transmission lines carry electric energy from one point to another in an electric power system. They can, as briefly discussed earlier, carry AC or direct current (DC), or a combination of both. In addition, electric current may be carried by either overhead or underground lines. Figure 2–3 shows overhead power transmission lines crossing the San Fernando Valley, while figure 2–4 illustrates underground power lines.

Fig. 2–3. Transmission lines crossing the San Fernando Valley

Fig. 2–4. Transmission line laid in a trench

The main distinguishing characteristics between transmission lines and distribution lines are that transmission lines are operated at relatively high voltages, they transmit large quantities of power, and they transmit power over long distances. Figure 2–5 shows some typical transmission structures. Distribution, which will be discussed more fully later in this chapter, generally includes medium-voltage (less than 50 kV) power lines, low-voltage electrical substations and pole-mounted transformers, low-voltage (less than 1,000 kV) distribution wiring, and sometimes electricity meters.

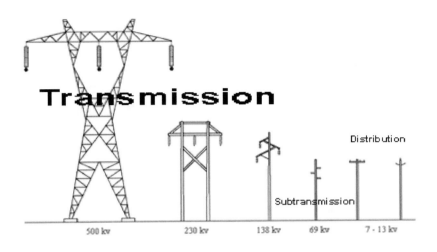

Fig. 2–5. Typical transmission structures

There are three basic types of transmission lines:

- Overhead transmission lines
- Subtransmission lines
- Underground transmission lines

Overhead AC transmission lines carry three-phase current. The voltages vary according to the particular power grid system to which they belong. Transmission voltages can vary from 69 kV up to 765 kV. Figure 2–6 shows that the DC voltage transmission tower has lines in pairs rather than in threes (for three-phase current) as in AC voltage lines. One line is the positive current line and the other is the negative current line.

Subtransmission lines carry voltages reduced from the major transmission line system. Typically 34.5 kV to 69 kV, this power is sent to regional distribution substations. Sometimes the subtransmission voltage is tapped along the way for use in industrial or large commercial operations. Some utilities categorize these as transmission lines. Figure 2–7 depicts the relative placement of subtransmission lines in relation to distribution lines. Figure 2–8 gives a view of subtransmission lines with distribution primaries and secondaries.

Fig. 2–6. DC voltage transmission lines

Fig. 2–7. Transmission lines are on the top, while subtransmission lines are on the bottom

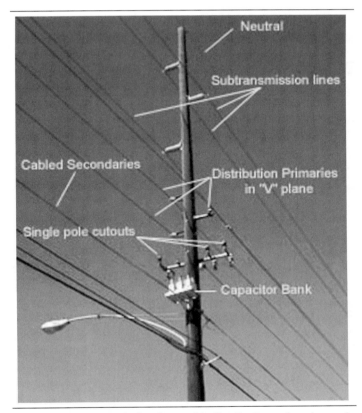

Fig. 2–8. Subtransmission lines with distribution primaries and secondaries

Underground transmission lines are often used in highly populated areas. They also are commonly used for aesthetics where homeowners, landowners, or business owners do not wish to have overhead transmission lines. Figure 2–9 illustrates a cross section of an underground transmission line. Underground transmission lines are most usually placed in conduits, trenches, or tunnels.

Conductor Size : 2500mm²
Diameter : 170mm
Weight : 43kg/m

Fig. 2–9. Cross section of underground transmission line

The role of substations

By definition, substations are high-voltage electric system facilities. Substations play many roles in the ultimate delivery of power to end users. These could include switching generators, equipment, and circuits or lines in and out of a system, changing AC voltages from one level to another, and/or changing AC to DC or DC to AC.

Substation sizes vary. Some may be small, with little more than a transformer and associated switches. Other substations are grand in scale, featuring several transformers and dozens of switches and other equipment. Figure 2–10 depicts a typical substation and the functions the system performs. Although not all substations are designed to function exactly alike, there are some functions performed at most substations:

- Change voltage from one level to another
- Regulate voltage to compensate for system voltage changes
- Switch transmission and distribution circuits into and out of the grid system
- Measure electric power qualities flowing in the circuits
- Connect communication signals to the circuits
- Eliminate lightning and other electrical surges from the system
- Connect electric generation plants to the system
- Make interconnections between the electric systems of more than one utility
- Control reactive kilovolt-amperes supplied and the flow of reactive kilovolt-amperes in the circuits

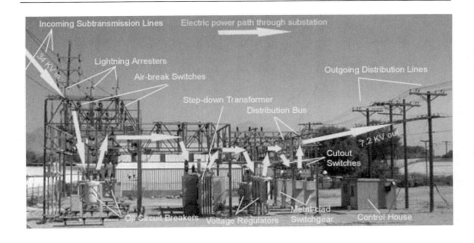

Fig. 2–10. Typical substation

While there are generally four types of substations, there are some substations that are a combination of two or more types. Substations types include:

1. Step-up transmission substation

2. Step-down transmission substation

3. Distribution substation

4. Underground distribution substation

A step-up transmission substation receives electric power from a nearby generating facility and uses a large power transformer to increase the voltage for transmission to distant locations. A transmission bus is used to distribute electric power to one or more transmission lines. There can also be a tap on the incoming power feed from the generation plant to provide electric power to operate equipment in the generation plant.

A substation can have circuit breakers that are used to switch generation and transmission circuits in and out of service as needed or for emergencies requiring the shutdown of power to a circuit or redirection of power.

The specific voltages leaving a step-up transformer substation are determined by the customer needs of the utility supplying the power and the requirements of any connections to regional grids. Table 2–1 shows typical voltages. As shown in table 2–1, DC voltage is either positive or negative polarity. A DC line has two conductors, so one would be positive and the other negative.

Table 2–1. Typical voltages

Voltage Type	Transmission Voltages
High Voltage AC	69kV, 115kV, 138kV, 161kV, 230kV
Extra-High Voltage AC	345 kV, 500kV, 765kV
Ultra-High Voltage AC	1100kV, 1500kV
DC High-Voltage	+/-250kV, +/-400kV, +/-500kV

Source: OSHA

Step-down transmission substations are located at switching points in an electrical grid. They connect different parts of a grid and are a source for subtransmission lines or distribution lines. The step-down substation can change the transmission voltage to a subtransmission voltage, usually 69 kV. The subtransmission voltage lines can then serve as a source to distribution substations. In some cases, power is tapped from the subtransmission line for use in an industrial facility along the way. Otherwise, the power goes to a distribution substation.

Distribution substations are located near end users. Three-phase systems can be connected in two different ways. If the three common ends of each phase are connected at a common point and the other three ends are connected to a three-phase line, it is called a *wye*, or *Y* connection. If the three phases are connected in a series to form a closed loop, it is called a *delta* connection. Distribution substation transformers change the transmission or subtransmission voltage to lower levels for end-user consumption. Typical distribution voltages vary from 34,500Y/19,920 V to 4,160Y/2,400 V. A three-phase circuit with a grounded neutral source is interpreted as 34,500Y/19,920 V. This would have three high-voltage conductors or wires and one grounded neutral conductor (a total of four wires). The voltage between the three-phase conductors or wires would be 34,500 V. The voltage between one phase conductor and the neutral ground would be 19,920 V. From here, the power is then distributed to industrial, commercial, and residential customers.

Underground distribution substations are also located near end users. Distribution substation transformers change the subtransmission voltage to lower levels for end-user consumption. Typical distribution voltages vary from 34,500Y/19,920 volts to 4,160Y/2,400 volts.

As shown in figure 2–11, an underground system may consist of many parts:

- Conduits
- Duct runs
- Manholes
- High-voltage underground cables
- Transformer vault
- Riser
- Transformers

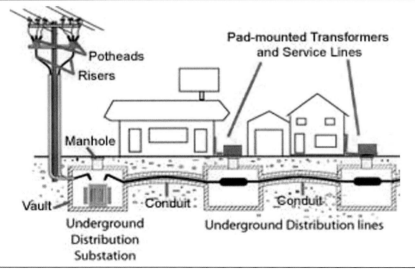

Fig. 2–11. Underground distribution substation

Of special note are underground transformers. They are essentially the same as aboveground transformers, but are constructed for the particular needs of underground installation. Vault type, pad-mounted, submersible, and direct-buried transformers are all used in underground systems. Perhaps two of the most common ones, vault type and pad mounted, are shown in figure 2–12 and figure 2–13, respectively.

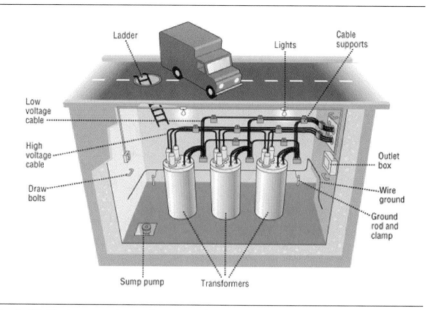

Fig. 2–12. Underground transformer vault

Fig. 2–13. Pad-mounted underground transformer

Another part of special importance is the riser. The riser is a set of devices connecting an overhead line to an underground line. Figure 2–14 illustrates how a riser has a conduit from the ground up the pole where potheads are use to connect to the overhead lines.

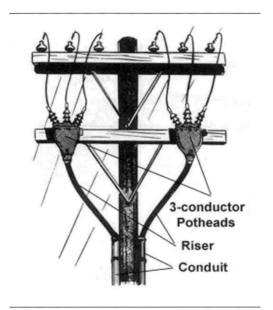

Fig. 2–14. Riser diagram

Distribution systems

A distribution system originates at a distribution substation and includes the lines, poles, transformers, and other equipment needed to deliver electric power to consumers at the required voltages. Figure 2–15 shows some of the major equipment on a power pole needed to deliver electricity.

Required voltages will depend upon the ultimate customer. Customers may be classed as:

- Industrial
- Commercial
- Residential
- Transportation

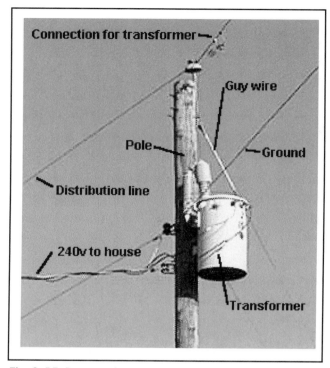

Fig. 2–15. Power pole parts

Most industrial customers need anywhere from 2,400 V to 4,160 V to run heavy machinery. They often have their own substation to reduce the voltage from the transmission line to the desired level for distribution throughout the plant area. Industrial customers usually require three-phase lines to power three-phase motors.

Commercial customers are usually served at distribution voltages. These range from 14.4 kV to 7.2 kV through a service drop line, which leads from a transformer on or near the distribution pole to the customer's end-use structure. Commercial end users may require three-phase lines to power three-phase motors.

For residential consumer use, distribution electricity is reduced to the end-use voltage (120/240 V single phase) via a pole-mounted or pad-mounted transformer (discussed earlier). Power is delivered to the residential customer through a service drop line, which leads from the distribution pole transformer to the customer's structure, for either overhead lines or underground lines.

Light rail and subway systems are currently the only electric transportation systems. A small distribution substation reduces the local distribution voltage to the transportation system requirements. The overhead lines supply electric power to the transportation system motors, and the return current lines are connected to the train tracks.

As touched on above, a distribution system consists of all the facilities and equipment connecting a transmission system to the customer's equipment. There are many parts to a typical distribution system:

- Substations
- Distribution feeder circuits
- Switches
- Protective equipment
- Primary circuits
- Distribution transformers
- Secondaries
- Services

Substations, as discussed earlier, are high-voltage electric system facilities. They switch generators, equipment, and circuits or lines in and out of a system. In addition, substations change AC voltages from one level to another, and/or change AC to DC or DC to AC.

Distribution feeder circuits are the connections between the output terminals of a distribution substation and the input terminals of primary circuits. The distribution feeder circuit conductors leave the substation from a circuit breaker or circuit recloser via underground cables, called substation exit cables. The underground cables connect to a nearby overhead primary circuit outside the substation. This eliminates multiple circuits on the poles adjacent to the substations.

Several distribution feeder circuits can leave a substation, extending in different directions to serve customers. The underground cables are connected to the primary circuit via a nearby riser pole. The distribution feeder bay routes power from the substation to the distribution primary feeder circuits.

Distribution systems have switches installed at strategic locations to redirect or cut off power flows for load balancing or sectionalizing. Also, this permits repairing of damaged lines or equipment or upgrading work on the system. There are many types of switches:

- Circuit-breaker switches
- Single-pole disconnect switches
- Three-pole, group-operated switches
- Pad-mounted switchgear

Protective equipment in a distribution system consists of protective relays, cutout switches, disconnect switches, lightning arresters, and fuses. All these parts work individually or in concert with one another to open circuits whenever a short circuit, lightning strike, or other disruptive event occurs.

When a circuit breaker opens, the entire distribution circuit is deenergized. Since this can disrupt power to many customers, the distribuiton system is often designed with many layers of redundancy. Through redundancy, power can be shut off in portions of the system only, but not the entire system, or can be redirected to continue to serve customers. Only in extreme events, or failure of redundant systems, does an entire system become deenergized, shutting off power to large numbers of customers.

The redundancy consists of the many fuses and fused cutouts throughout the system that can disable parts of the system but not the entire system. Lightning arresters, as shown in figure 2–16, also act locally to drain off electrical energy from a lightning strike so that the larger circuit breakers are not actuated.

Primary curcuits are the distribution circuits that carry power from substations to local load areas. They are also called express feeders or distribution main feeders. The distribution feeder bay routes power from the substation to the distribution primary feeder circuits.

Distribution transformers reduce the voltage of the primary circuit to the voltage required by customers. This voltage varies and is usually:

- 120/240 V single-phase for residential customers
- 480Y/277 or 208Y/120 for commercial or light industry customers

Fig. 2–16. Substation bus lightning arresters

Three-phase pad-mounted transformers are used with an underground primary circuit and three single-phase pole-type transformers for overhead service. Network service can be provided for areas with large concentrations of businesses. These are usually transformers installed in an underground vault. Power is then sent via underground cables to separate customers.

Secondaries are the conductors originating at the low-voltage secondary winding of a distribution transformer. Secondaries are three-wire, single-phase circuits. They extend along the rear lot lines, alleys, or streets past customers' premises. Overhead secondary lines are normally strung below the primary lines, typically in a vertical plane. When secondaries are in a vertical plane, they are directly attached to the support pole, one above the other.

Wires extending from the secondaries or distribution transformer to a customer's location are called a service. A service can be above ground or below ground. Underground services have a riser connection at the distribution pole. Commercial and residential services are much the same and can be either 120 V or 220 V, or both.

New Technologies

According to the DOE, power electronics communication devices hold much promise for transforming the electric power system. High-voltage power electronics allow precise and rapid switching of electric power to support long-distance transmission. On the other hand, lower voltage power electronics can be used in power distribution, and in the interface between customers and the electric grid.

While power electronics are at the heart of the interface among energy storage, distributed generation, and the electric system, high cost and a lack of proven performance information remain significant barriers to its expanded use. Questions surrounding reliability and durability over a period of time in real-world applications continue to cloud the issue of whether to roll out power electronics or continue to sit back and wait.

One core capability of power electronics, however, may present enough opportunities for expanding functionality and improving system operations that its use may be compelled despite surrounding questions. While power electronics devices can be applied in place of traditional power devices such as switches, controllers, capacitors, and condensors, they have the capability to perform several of these functions within a single device.

Technology needs for power electronics include both short- and long-term opportunities. Over the next 5 to 20 years, the DOE predicts that advances in power electronics technologies could revolutionize many aspects of power system operations and planning. This includes the expanded use of DC for both transmission and distribution. While long-term opportunities are substantial, there are near-term opportunities to explore even more.

This is where the DOE's GridWorks program comes into play. A new program activity in the DOE's Office of Electric Transmission and Distribution (OETD), GridWorks has the goal of improving electric system reliability through the modernization of key grid components. These include cables and conductors, substations and protective systems, and power electronics.

Fiscal year 2005 was the first year funds were appropriated to support GridWorks activities. In consultation with electric utility industry representatives, power system equipment manufacturers, other federal and state agencies, universities, and national laboratories, GridWorks is ready to incrementally improve existing power systems. It is also prepared to accelerate their introduction into the marketplace. Other activities on the board are to develop new technologies, tools, and techniques to support the modernization of the electric grid for 21st century requirements.

New power grid concepts

A variety of approaches exist for improving electricity grid operation. According to DOE, all of the approaches are motivated by power reliability and quality concerns, and all of them incorporate distributed energy (DE).

Some of the more popularly noted power grid concepts include:

- Minigrids
- Power parks
- DC microgrids
- Flexible AC transmission systems (FACTS)
- Electric load as a reliability resource

Minigrids are a set of generators and load-reduction technologies that supply the entire electricity demand of a localized group of customers. A minigrid, also known as a microgrid, can significantly improve the economics of meeting energy needs using DE. This is accomplished by avoiding the cost of transmitting electricity from a distant central station power or transporting fuel from a distant supply source.

Using the same or similar technologies used by electric utilities in distributed power applications, minigrids are different in that they are not always connected to the central grid. In some cases, the generators and other distributed resources are installed to relieve utility constraints on the existing grid. The intent is to possibly disconnect these generators and their load from the grid at a later point in time.

In other cases, an electrically isolated minigrid can be created. If the option becomes available and is desirable, this minigrid can be integrated with the central grid. Thus, using a mix of generating and demand-side management (DSM) technologies gives the power supplier flexibility to meet a wider range of loads.

Power parks offer up an alternative to traditional grid technologies to achieve reliability. The traditional approach has been to supply multiple power feeders to the system and provide a backup line from a hydropower station, for example. This is both expensive in terms of dollars and public relations (guaranteeing reliability at the expense of service to other customers).

Power parks, which typically include an on-site power source to increase reliability, include uninterruptible power supplies such as battery banks, ultracapacitors, or flywheels. One of the earliest power parks, built in 1998, is the PEI Power Park in Archibald, Pennsylvania. A CHP plant there provides both steam and electricity to occupants of the park.

In May 2000, the University Research Park adjacent to the University of California Irvine campus was designated by the DOE as a power park. It develops a number of DE resources, including fuel cells, gas turbine engines, microturbine generators, and photovoltaics, to meet the park's power needs. Other parks are either in the building process or are up and running. One newer one is a park being built by Hunt Power in McAllen and Mission, Texas. This park features options for reliable on-site power generation and redundant telecommunications feeds.

DC microgrids would allow neighborhoods to run entirely on DC. A high-voltage DC line would interface with the rest of the grid through high-tech DC-to-AC converters. DC systems are less vulnerable to power quality issues, and digital devices run on DC current. DC systems also allow distributed generation equipment to be connected directly with the microgrid without using DC-to-AC converters at the power source. The good news is that the converter technology needed to interface these DC microgrids with the AC power grid is proving, thus far, to be cost-effective.

FACTS is yet another concept for future power grids. FACTS incorporates high-current and high-voltage power electronic devices to increase the carrying capacity of individual transmission lines and improve overall system reliability by reacting very quickly to grid disturbances.

Utilizing the responsive electronic devices, the electric power industry envisions converting the electric power grid to more a of networked system. This system would respond in real time to a broader dispersion of electric generators, higher and less predictable line loadings, and a vast increase in transactions. The data and control system needed to achieve such a system would likely be a dispersed network, much like the Internet. DOE maintains that there is at least one technical problem of achieving this system—how to decentralize the control of the system while maintaining the essential balance between electrical loads and electrical generation.

Two types of electrical load are also a valuable reliability resource:

- Price-responsive load
- Emergency-responsive load

Price-responsive load, which requires real-time pricing, puts the electric customer in the driver's seat. Customers see the fluctuating prices of energy during the day, and then choose whether or not to buy at a given price. When generation is in short supply and prices rise, more and more customers with accurate pricing information are likely to turn off loads. At some point, load will match generation without having to resort to voltage reductions or to rolling blackouts.

In order for price-responsive load to operate, customers must be provided with price signals, preferably on an automatic basis.

Emergency-responsive load is commonly thought of as a reliability resource. It requires a customer to bid load into the operational reserve market for a given day and, if called upon, to turn the load off within 10 minutes if the load was specified as a spinning reserve. The customer must turn off the load within a half hour if the load was specified as a supplemental reserve. The customer is paid for the reserve capacity whether or not it is called upon.

For emergency-responsive load, some form of load aggregation and a means of certifying the load was available to turn off and was turned off when requested may be needed. Some loads, such as water heating, can respond quickly to serve as spinning reserve. Other loads may not be able to respond as quickly, but may still qualify as supplemental reserves that can be dropped within a half hour.

Both price- and emergency-responsive loads will require markets to be designed to give the customer correct signals for participation in competitive energy and/or ancillary services markets in an efficient manner. Thus electrical load as a reliability resource may indeed be difficult to implement for the aforementioned reason and because of the advanced technological and procedural innovation it will demand.

For example, when the supply of electricity is insufficient to meet overall demand, the price of wholesale power from electricity generators goes up. This fact has largely been hidden from most electricity customers, who typically have paid their utilities a fixed rate for each unit of electricity consumed, regardless of the time in which it was used. Real-time pricing, by its operational definition, means passing fluctuations in the true cost of electricity on to customers, so they have the pricing information they will need to adjust their electricity usage.

In addition, this resource will mean new communication and control technologies must be employed. Electricity system operators would require more precise information about demand fluctuations. Operational control systems that can respond to load reductions on par with power generation will need to be given high priority should electrical load become a reliability resource.

And, finally, some electrical equipment, such as induction motors and various other power electronics devices, may create some challenges for reliable grid operation. These types of equipment have tremendous impact on the grid. For price- or emergency-responsive loads, this equipment may need to be redesigned or operated in such a way to reduce its effects on the grid.

The challenge ahead

According to the Electric Power Research Institute (EPRI), the energy industry continues in its pattern of globalizaion, disaggregation, deregulation, and restructuring. As it does so, traditional and emerging transmission and distribution companies are faced with critical decisions that will impact their survival.

Open access markets, the saturation of the exisiting transmission grid, and the emergence of distributed resource technologies are forcing power delivery entitites to deal with significant and often unpredictable complexity. The greatest challenge facing transmission and distribution companies, EPRI contends, centers on enhancing system reliability and performance while maximizing the utilization of power delivery system assets.

Providing the major backbones of the energy industry, transmission companies are subject to the same uncertainties and opportunities as the industry at large. At the federal level, for instance, the three major interconnected grids in the United States have been deregulated to provide open and competitive access. The traditional and emerging owners and operators of these networks are being confronted with power flows, patterns, and magnitudes not contemplated in the original planning, design, and construction of transmission facilities.

As well, distribution companies are at a pivotal position in the energy industry from both a profit and strategic viewpoint. States are moving at widely varying speeds toward independent visions of tomorrow's electrical distribution company. However, the competitive marketplace requires that distribution companies identify and implement new technologies, tools, and methods to enhace levels of reliability, performance, and service. For some distribution companies, the future may strictly be a wires-only commodity focus. For others, a strategy of targeting end-use customers with a portfolio of innovative products and services may be employed.

The choices of which strategies and tactics to undertake are thus critical. Distribution companies will need to take prudent measures to ensure that the distribution system has enhanced reliability and performance. Power delivery entities will need to be aware of the highly technologically dependent business environment. They must remain keenly aware that system disturbances and outages have significant financial, political, legal, as well as strategic positioning and branding impacts.

Part II:
The Formation and
Reformation of the
Electric Utility Industry

The Beginning
of the
Electric Utility Industry

The rich and long history of the making of the electric utility industry is fascinating. It is also a biographical essay in the understanding and appreciation of the men behind the scenes, as well as the technologies they so painstakingly developed. It is because of a handful of inventors/scientists/ engineers that this "industry" even exists.

In this chapter, early history and the men who made history come to the forefront. Their efforts foretell the inventive shaping of the current electric utility industry, including its machinery, science, logic, and research behind the machinery and operations, and even its economic aspects and impacts.

The Early Years

Very early history of electricity dates back to Greek philosopher Thales of Miletus, who lived around 600 B.C. This philosopher was likely the man who discovered that amber acquires the power to attract light objects when rubbed. Another Greek philosopher, Theophrastus, in a treatise written about three centuries later, stated that this power is possessed by other substances.

The first scientific study of electrical and magnetic phenomena, however, did not appear until 1600 A.D., when the research of English physician William Gilbert was published. Gilbert was the first to apply the term "electric" (Greek *electron*, "amber") to the force that substances exert after rubbing. He also distinguished between magnetic and electric action.

The first machine for producing an electric charge was described in 1672 by the German physicist Otto von Guericke. It consisted of a sulfur sphere turned by a crank on which a charge was induced when the hand was held against it. The French scientist Charles François de Cisternay Du Fay was the first to make clear the two different types of electric charge—positive and negative.

The earliest form of condenser, the Leyden jar, was developed in 1745. It consisted of a glass bottle with separate coatings of tin foil on the inside and outside. If either tin foil coating was charged from an electrostatic machine, a violent shock could be obtained by touching the foil coatings at the same time.

Benjamin Franklin spent much of his time in electrical research. His famous kite experiment proved that the atmospheric electricity that causes the phenomena of lightning and thunder is identical with the electrostatic charge on a Leyden jar. Franklin developed a theory that electricity is a single "fluid" existing in all matter and that its effects could be explained by excesses and shortages of this fluid.

In 1766 British chemist Joseph Priestly proved experimentally the law that the force between electric charges varies inversely with the square of the distance between the charges. Priestly also demonstrated that an electric charge distributes itself uniformly over the surface of a hollow metal sphere, and that no charge and no electric field of force exist within such a sphere.

Charles Augustin de Coulomb invented a torsion balance to measure accurately the force exerted by electrical charges. With this apparatus, he confirmed Priestley's observations and showed that the force between two charges is also proportional to the product of the individual charges. Faraday, who made many contributions to the study of electricity in the early 19th century, was also responsible for the theory of electric lines of force.

Italian physicists Luigi Galvani and Alessandro Volta conducted the first important experiments in electrical currents. Galvani produced muscle contraction in the legs of frogs by applying an electric current to them. In 1800, Volta announced the first artificial electrochemical source of potential difference, a form of electric battery.

Danish scientist Hans Christian Oersted in 1819 demonstrated the fact that a magnetic field exists around an electric current flow. In 1831, Faraday proved that a current flowing in a coil of wire can induce electromagnetically a current in a nearby coil. Around 1840, James Prescott Joule and the German scientist Hermann von Helmholtz demonstrated that electric circuits obey the law of the conservation of energy and that electricity is a form of energy.

The work of British mathematical physicist James Clerk Maxwell made an important contribution to the study of electricity in the 19th century. Maxwell investigated the properties of electromagnetic waves and light, and developed the theory that the two are identical.

Maxwell's work paved the way for German physicist Heinrich Rudolf Hertz, who produced and detected electric waves in the atmosphere in 1886. It also paved the way for Italian engineer Guglielmo Marconi, who in 1896 harnessed these waves to produce the first practical radio signaling system.

The electron theory, which is the basis of modern electrical theory, was first advanced by the Dutch physicist Hendrik Antoon Lorentz in 1892. The charge on the electron was first accurately measured by the American physicist Robert Andrews Millikan in 1909. The widespread use of electricity as a source of power is largely due to the work of such pioneering engineers and inventors as Thomas Alva Edison, Nikola Tesla, and Charles Proteus Steinmetz. More about these talented gentlemen follows.

Major Players

There were many individuals who played direct and important roles in furthering the advancement of electricity into basic as well as luxury applications of the technology into daily life. These efforts ultimately led to the formation of the electric utility industry. However, the remainder of this chapter will focus on four of the most talked-about contributors to electricity and the industry itself. History has shown that the four scientists, inventors, and astute mathematicians of Thomas Alva Edison, Nikola Tesla, Charles Steinmetz, and Samuel Insull all played significant and highly noteworthy roles. It is for this reason that the remainder of this chapter will focus on their work.

Thomas Alva Edison

Edison is seen as perhaps one of the most prolific inventors of practical electrical devices in history. His inventions were so far-reaching that historians can with unfaltering certainty say Edison actually helped shape modern society. This is not so far-fetched, given that he developed a practical electric light bulb, electric generating system, sound-recording device, and the motion picture projector.

Edison, shown in his East Orange, New Jersey laboratory in figure 3–1, was born in Milan, Ohio on February 11, 1847. He attended school for only three months in Port Huron, Michigan. When he was just 12 years old, he began selling newspapers on the Grand Trunk Railway, devoting his spare time primarily to experimentation with printing presses and with electrical and mechanical apparatuses.

Fig. 3–1. Thomas Edison in his New Jersey laboratory, 1901

In 1862, Edison published a weekly newspaper, the **Grand Trunk Herald**, printing it in a freight car that doubled as his laboratory. For saving the life of a station official's child, he was rewarded with being taught telegraphy. While working as a telegraph operator, Edison made his first important invention, a telegraphic repeating instrument. This enabled messages to be transmitted automatically over a second line without the presence of an operator.

Next, Edison secured employment in Boston, devoting his spare time to research. He invented a vote recorder that, although possessing many merits, was not sufficiently practical to warrant its adoption. He also devised and partly completed a stock-quotation printer. Later, while employed by the Gold and Stock Telegraphy Co. of New York City, Edison greatly improved its apparatus and service. Edison earned $40,000 in sales of his telegraphic appliances. With this money, it is reported by historians, he established his own laboratory in 1876.

While working in his laboratory, Edison devised an automatic telegraph system, making greater speed and range of transmission a reality. Perhaps one of Edison's most far-reaching achievements in telegraphy was his invention of machines that made possible simultaneous transmission of several messages on one line, thus greatly increasing the usefulness of existing telegraph lines.

In 1877, Edison announced his invention of a phonograph, in which sound could be recorded mechanically on a tinfoil cylinder. Two years later in 1879, he publicly exhibited his incandescent electric light bulb. The electric light bulb would become Edison's most important invention, and the one requiring the most careful research and experimentation. This new source of light was a smashing success.

Building on this newfound success, Edison busied himself with the improvement of the electric light bulb, and of the dynamos for generating the necessary electric current. In 1882, he developed and installed the world's first large central electric power station in New York City. Edison's use of DC, however, lost out to the AC system developed by American inventors Nikola Tesla and George Westinghouse.

In 1887, Edison moved his laboratory from Menlo Park, New Jersey, to West Orange, New Jersey, where he would build a large laboratory for experimentation and research. Just one year later, he invented the kinetoscope, the first machine to produce motion pictures by a rapid succession of individual views.

Among Edison's noteworthy inventions in later years was the Edison storage battery (an alkaline, nickel-iron storage battery). The battery, the result of many thousands of experiments, was particularly rugged and had a high electrical capacity per unit of weight. He also developed a phonograph in which the sound was impressed on a disk instead of a cylinder. By synchronizing his phonography and kinetoscope, Edison produced the first talking moving pictures in 1913.

Other noteworthy Edison discoveries include the electric pen, the mimeograph, the microtasimeter (used for the detection of minute changes in temperature), and a wireless telegraphic method for communicating with moving trains. Edison died in West Orange on October 18, 1931, after having patented more than 1,000 inventions.

Nikola Tesla

Perhaps best known for his invention of the AC induction motor (see fig. 3–2), Nikola Tesla (1856–1943) was a Serbian-born American physicist, electrical engineer, and inventor. Tesla's inventions included a telephone repeater, rotating magnetic field principle, polyphase AC system, induction motor, AC power transmission, Tesla coil transformer, wireless communication, radio, fluorescent lights, and more than 700 other patents.

Fig. 3–2. Tesla's AC induction generator

Tesla, pictured in figure 3–3, was educated at the Polytechnic School in Graz, Austria, and at the University of Prague. While at Graz, he first saw the Gramme dynamo, which operated as a generator and, when reversed, became an electric motor. This was the point at which Tesla conceived a way to use AC to its best advantage.

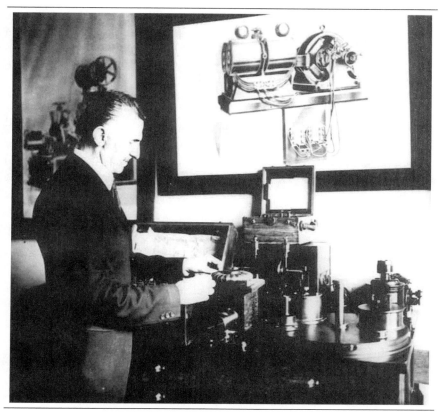

Fig. 3–3. Nikola Tesla

Later, at Budapest, Tesla came up with the principle of the rotating magnetic field and developed plans for an induction motor. This would be the first step toward his successful utilization of AC. In 1882, Tesla went to work in Paris for the Continental Edison Company, and while on assignment to Strasbourg in 1883, he constructed his first induction motor.

Tesla went to America in 1884 and first found employment with Thomas Edison. However, he soon parted ways with Edison. Historians cite that the two inventors were far apart in their backgrounds and methods and thus their separation became inevitable.

In May 1885, George Westinghouse, head of the Westinghouse Electric Company in Pittsburgh, bought the patent rights to Tesla's polyphase system of AC dynamos, transformers, and motors. This transaction would tip off the power struggle between Edison's DC systems and the Tesla-Westinghouse AC approach, which eventually won out.

In his 1938 speech before the Institute of Immigrant Welfare in New York, Tesla said, "George Westinghouse was, in my opinion, the only man on this globe who could take my alternating-current system under the circumstances then existing and win the battle against prejudice and money power. He was a pioneer of imposing stature, one of the world's true noblemen of whom America may well be proud and to whom humanity owes an immense debt of gratitude."

Tesla went on to establish his own laboratory, experimenting with shadowgraphs similar to those that later were to be used by Wilhelm Röntgen when he discovered X-rays in 1895. Tesla's other experiments included work on a carbon button lamp, the power of electrical resonance, and various types of lighting.

Exhibitions in his laboratory, in which Tesla lighted lamps without wires by allowing electricity to flow through his body to allay fears of AC, were quite common. The Tesla coil, which he invented in 1891, is widely used today in radio and television sets and other electronic equipment. This same year, Tesla obtained his U.S. citizenship.

In 1893, Westinghouse used Tesla's system to light the World's Columbian Exposition in Chicago. Westinghouse's success was a factor in winning him the contract to install the first power machinery at Niagara Falls, which bore Tesla's name and patent numbers. The project carried power to Buffalo, New York by 1896.

Charles Proteus Steinmetz

A German-American electrical engineer and inventor, Charles Proteus Steinmetz (1865–1923) was a pioneer in the field of electrical engineering. While born with a major physical deformity (hunchback), Steinmetz never let that stand in the way of his achievements. Much like Thomas Alva Edison, Steinmetz was also a prolific inventor and achieved nearly 200 patents for various electrical devices.

It is safe to say that without Steinmetz's development of AC theories, the expansion of the electric power industry in the United States in the early 20th century would have been impossible, or at least greatly delayed.

Among his greatest achievements was the invention of a commercially successful AC motor. Steinmetz had some notable accomplishments:

- His work in the field of electromagnetism
- The development of a practical, simplified method of managing and calculating values for AC using complex numbers
- His research on lightning phenomena
- His invention of the three-phase electrical circuit

Steinmetz, shown in figure 3–4, is worthy of mention here because, according to historians, his work made possible the expansion of the electric power industry in the United States. ("What Steinmetz Knew and What He Charged for It" offers a glimpse of Steinmetz's talent and his humor in the accompanying sidebar.)

Fig. 3–4. Charles Steinmetz in 1890

What Steinmetz Knew and Charged for It

The following is an anecdote, as told by Charles M. Vest, President of the Massachusetts Institute of Technology, during commencement on June 4, 1999:

In the early years of this century, Steinmetz was brought to General Electric's facilities in Schenectady, New York. GE had encountered a performance problem with one of their huge electrical generators and had been absolutely unable to correct it. Steinmetz, a genius in his understanding of electromagnetic phenomena, was brought in as a consultant—not a very common occurrence in those days, as it would be now. Steinmetz also found the problem difficult to diagnose, but for some days he closeted himself with the generator, its engineering drawings, paper and pencil. At the end of this period, he emerged, confident that he knew how to correct the problem. After he departed, GE's engineers found a large 'X' marked with chalk on the side of the generator casing. There also was a note instructing them to cut the casing open at that location and remove so many turns of wire from the stator. The generator would then function properly. And indeed it did. Steinmetz was asked what his fee would be. Having no idea in the world what was appropriate, he replied with the absolutely unheard of answer that his fee was $1,000. Stunned, the GE bureaucracy then required him to submit a formally itemized invoice. They soon received it. It included two items: 1. Marking chalk 'X' on the side of generator: $1, and 2. Knowing where to mark chalk 'X': $999."

In 1889, Steinmetz established a small laboratory at a factory in Yonkers, New York, under the tutelage of his employer, Rudolph Eickemeyer. Eickemeyer had invented hat-making machinery and wanted to expand into electrical motors and generators, a brand new field at that time.

Steinmetz' experiments on power losses in the magnetic materials used in electrical machinery led to his first important work, the law of hysteresis. This law deals with the power loss that occurs in all electrical devices when magnetic action is converted to unusable heat. Until that time, the power losses in motors, generators, transformers, and other electrically powered machines could be known only after they were built.

Once Steinmetz had discovered the law governing hysteresis loss, engineers could calculate and minimize losses of electric power due to magnetism in their designs before starting the construction of such machines.

Steinmetz' second contribution was a practical method for making calculations concerning AC circuits. This method was an example of using mathematical aids for engineering the design of machinery and power lines. Its value was such that the performance of the electrical system could be predicted in advance without the necessity of going through the expensive and uncertain process of building the system first and then testing its efficiency. Steinmetz developed a symbolic method of calculating AC phenomena and in so doing, simplified an extremely complicated and barely understood field so that even the most average engineer could work with AC. This accomplishment was largely responsible for the rapid progress made in the commercial introduction of AC apparatus.

In 1886, Thomas Edison founded the General Electric Company and wanted to hire Steinmetz. In 1893, the newly formed General Electric Company purchased Eickemeyer's company, primarily for his patents, but Steinmetz was considered one of its major assets. In 1894, Steinmetz was transferred to the main General Electric plant at Schenectady, New York. His original residence in Schenectady is still standing at 53 Washington Street.

While at General Electric, Steinmetz gained an expanded opportunity for research and implementation of his ideas. He was assigned to the new calculating department. His first task was to work on the company's proposal for building the generators at the new Niagara Falls power station. While in this position, Steinmetz began indoctrinating engineers with his method of calculating AC circuits. He would work the next 20 years of his life preparing a series of masterful papers and books to further his profound knowledge of AC circuits.

In 1903, Steinmetz served as a professor of electrical engineering and applied physics at Union College in Schenectady. He served in this capacity until 1913, during which time historians contend he guided the college in becoming the best in the nation. Steinmetz would never accept remuneration for his services as professor, and according to the Institute of Chemistry at The Hebrew University of Jerusalem, he said his only wish for students was "the spirit of divine discontent, for without it the world would stand still."

Steinmetz' third major scientific achievement was in the study and theory of electrical transients or changes in electrical circuits of short duration.

A prime example of this would be lightning. Steinmetz' investigation of lightning phenomena resulted in his theory of traveling waves and opened the way for his development of devices to protect high-power transmission lines from lightning bolts.

In the course of his work, Steinmetz also designed a generator that produced a discharge of 10,000 A and more than 100,000 V, equivalent to a power of more than 1,000,000 horsepower (hp) for 1/100,000 of a second.

As if all these accomplishments were not enough, Steinmetz in 1920 formed the Steinmetz Electric Motor Car Co. to design prototypes of several electric vehicles. The company, based in Brooklyn, New York, produced an industrial truck and a lightweight delivery car. His first electrical truck was churned out in 1922. Steinmetz had planned for his company to turn out 1,000 trucks and 300 cars annually, but that was cut short by his death in 1923. His company ceased operations shortly after his death.

Also according to the Institute of Chemistry, a friend close to him said, "Chapters have been written of his greatness intellectually; as many more could be filled with his kindnesses. Dwarfed, perhaps, in body, but with a heart as big as the universe and a soul as pure as a child's."

Samuel Insull

While there were many scientists, inventors, and engineers who helped pave the way to modern-day electric utility operations, Samuel Insull would play a major role in the economics of utility operations.

Insull, an ex-Edison lieutenant, became president of the Chicago Edison Company in 1892. Once in this position, he quickly learned that electric companies suffered high fixed costs associated with investments in generating plants and transmission equipment. He also discovered that operating costs were quite low. Insull proposed that with more customers on a system, more revenue could be generated, spreading out the utility's fixed costs.

To test this theory, Insull reduced electricity prices and aggressively marketed its benefits to attract more customers. During this time many electric utilities gave away light bulbs and electric irons to try to get more customers on their system.

Insull also discovered that the more time a generating plant was in use, the greater the efficiency factor (load diversity). This efficiency not only yielded higher profits, but lowered costs per kilowatt-hour to customers. Load diversity was one of Insull's most brilliant discoveries.

He would also find that there was a way of increasing efficiency through economies of scale. This meant using a single plant to service the morning and afternoon streetcar load, the daytime industrial load, and the evening residential load was much more economical than using three separate plants.

Insull welcomed the development of the demand meter, which made it possible for him to more accurately price the electricity his company sold to consumers. The demand meter was invented by Arthur Wright, an Englishman who worked at the Brighton, England municipal power company.

The new demand meter measured a customer's electric demand and the actual energy used. Insull set the price of electricity to cover both fixed costs (demand costs) and operating costs (energy or variable costs). Fixed costs refer to the fixed amount of investment that must be paid regardless of output, including power plant construction and equipment. Variable costs are those that vary with the level of electrical output and include fuel expenses.

The foregoing concepts, developed under Insull, would become important economic concepts, which to this day still help govern modern utility planning and pricing. These economic concepts can be recapped here:

- The more customers on a system, the more revenue can be earned.
- Load diversity as a means of yielding both higher profits and lower costs to customers.
- The economies of scale realized through the single plant concept.

In chapter 4, more history is discussed. In chapter 3, the focus was on the human factor in the shaping of the electric utility industry. In contrast, chapter 4 focuses on the early years of utility regulation, along with the various forms of regulation that remain applicable to electric utility companies.

The Electric Utility Industry as a Regulated Entity

4

Regulation brings to mind the terms "rules," "strictures," and even "accountability." While present day regulation of electric utilities has changed significantly since the early years of often inefficient and inflexible forms of oversight, the electric utility industry remains a regulated industry. This is true even despite the fact that competitive forces and a restructuring of the industry remain, and they likely will remain an undercurrent as well as a challenge of titanic proportions for utilities to navigate.

The Early Years of Regulation

Three different forms of federal government regulation were the rule of day in the early years:

1. Judicial regulation
2. Legislative regulation
3. Franchise regulation

Judicial regulation

Under judicial regulation, it was the norm for common law to develop from one individual lawsuit to the next. There were at least two major limitations. First, judicial regulation was costly. In certain instances, even though someone may have been injured, the issue may never have been made public due to the injured person's inability to afford judicial review of the harm.

Further, courts were charged by some with being severely deficient in retaining trained accountants and other experts to deal with the complex issues of the electric utility industry.

Another major limitation was that courts could not take legislative action. An example of this would be setting and enforcing new rates. Courts were only capable of ruling on matters at hand; their role was not to set matters right for the future. It would become obvious that direct legislative action would be needed to address these and other major limitations of the courts.

There were two principles or standards of common law that emerged out of judicial regulation:

1. **Monopoly and restraint of trade.** Both were considered to be contrary to the public interest. Because of this fact, lawsuits brought upon the standard of injury for either monopoly or restraint of trade were considered to be legitimate. In addition, it was likely that if the claims were proven, the offended party or parties could receive a damage award. (One may refer to the section on antitrust, which is an example of a modern-day parallel to this traditional common-law standard.)

2. **Common callings businesses.** Certain businesses were bound by the general rule that they were forbidden to refuse to sell a product or service to anyone. The common law standard set forth held that these types of businesses would serve all consumers at a reasonable cost and without any discrimination. (One may refer to the section on public utilities for a modern-day parallel to this standard.)

Legislative regulation

This form of regulation, while some improvement over judicial regulation, would also have limitations of its own. In the early 19th century, public utilities gained their incorporation status through their local legislative bodies. These legislative bodies granted charters, which included both corporate rights and some other special rights, such as the power of eminent domain.

Unlike the courts, early legislative bodies set maximum rates that electric utilities could charge their customers, and in some cases, legislatures held common stock yield to particular percentages.

In the late 19th century, states began to enact general incorporation laws. In the regulatory sense, incorporation laws were no more specific than the charters that were dispensed through the local legislative body.

With economic conditions changing at a rapid pace, legislative bodies, which often lacked the expertise to govern public utilities, proved to be an increasingly ineffective way to regulate the industry. Every change in the economy would precipitate correspondent changes to laws governing electric utilities. This process was time-consuming and made it difficult, if not impossible, for state legislatures to keep up with regulatory changes that needed to be made.

Franchise regulation

Franchise regulation was the direct precursor to state regulation of the electric utility industry. While city ordinances permitted a degree of local regulatory control over public utilities, it was franchise regulation that would become the center around which local regulation revolved.

In order for certain businesses to enter into and operate a new business under franchise regulation, business owners had to acquire a franchise from the relevant city council. In some situations the franchises were drafted in complete form. This included precise service standards, rates, applicable accounting methods, and in some cases, franchise renewal methods, which would all be spelled out under the terms of the franchise agreement with the business.

The downside was that franchises were often handed out surreptitiously by local city councils and were not granted for the life of the utility plant. Rather, the franchise agreement was only for a flexible duration, at the whim of the city councils. It was common for city council members to accept bribes and payoffs from utilities.

Exacerbating the problem was the fact that in the early years of the industry, a considerable amount of uneasiness surrounded the building, maintaining, and operating of power plants. Investor confidence was a challenge for utilities, since typical utility investors did not reap substantial monetary rewards for their investment.

It should come as little surprise then that franchise regulation grew to become an inefficient method of controlling public utilities. This was particularly so given the number and scope of changes taking place in public utility operations.

In the beginning, utilities served only one particular area (franchise area), but as time went on, technological advancements made it possible for one utility company to serve more than one area. This was due in large part to the economies of scale that could be produced when one, rather than several, public utilities could serve a mass of consumers. One utility serving many consumers also made for good economics, cutting out many of the inefficiencies and costs associated with such components as redundant equipment and transmission lines. It soon became clear that another type of regulation was needed.

While franchise regulation is not the norm, there are some cities that still issue franchises. For example, in Texas, incorporated cities are permitted to control rates and the services of electric, gas, and private water utilities within their boundaries. The public utility commission (PUC), in this instance, exercises appellate jurisdiction over electric and water rates within municipalities, and the Texas Railroad Commission exercises appellate jurisdiction over gas rates within local areas. Another example is found in Nebraska, where municipalities grant permits and also set rates for gas utilities.

State regulation

The concept of nonpartisan state agencies developed in the mid-1800s. Nonpartisan state agencies, it was thought, could better and more fairly monitor the franchise-granting process and the rates utilities set. Financing, as well, could be made easier and cheaper with these new state agencies at the helm.

While some early state commissions were created before 1870, they were largely fact-finding and advisory entities and only had jurisdiction over the railroads. State PUCs would be rather slow to develop. Generally, most nontransportation industries would not become subject to commission regulation until the beginning of the 20th century.

Samuel Insull was one of the more notable individuals to get state regulation for electric utilities off the ground. In 1898, Insull went before the National Electric Light Association (NELA) and proposed that state agencies be formed. (The NELA was the forerunner of the EEI, which represents the interests of electric IOUs.) The state agency proposal grew to be widely accepted in the industry during a time when there was an upsurge in the number of municipally owned systems. In fact, the number of municipally owned systems tripled between 1896 and 1906.

The state commission proposal was especially appealing for IOUs, due to the phenomenal growth in municipally owned systems. IOUs presumed they might gain more support from the public if privately held utility companies were regulated. This regulation, IOUs postulated, would serve to make the public feel more protected. In 1907, 3 states established state PUCs. By 1916, 33 states had formed them. Today, most states, plus the District of Columbia, have PUCs.

State Public Utility Commissions

Most state PUCs have either three or five members. There are a few states that have seven-member commissions. States with seven members on their commissions are Illinois, New York, North Carolina, and South Carolina.

In most cases, commissioners have had some experience in dealing with electric utility matters. The legal profession is the most common among commissioners. Each state has its own matrix of qualifications its commissioners must meet, such as age and professional requirements. Appointed by each state's governor or in some cases by the mayor, the state's legislature, or by popular vote, commissioners serve terms from between four and eight years, depending upon the state. Each PUC has an office staff that varies in size according to the number of members within the commission.

The National Association of Regulatory Utility Commissioners (NARUC) was set up as a quasi-judicial governmental not-for-profit corporation in 1889 to help with a variety of matters pertaining to the utility industry. For instance, NARUC has 16 standing committees:

1. Ad Hoc Committee on Critical Infrastructure

2. Ad Hoc Committee on Global Climate Change

3. Committee on Consumer Affairs

4. Committee on Electricity

5. Committee on Energy Resources and the Environment

6. Committee on Gas

7. Committee on International Relations

8. Committee on Telecommunications

9. Committee on Water

10. National Conference of State Transportation Specialists

11. Staff Subcommittee on Accounting and Finance

12. Staff Subcommittee on Electric Reliability

13. Staff Subcommittee on Executive Management

14. Staff Subcommittee on Information Services

15. Subcommittee on Nuclear Issues—Waste Disposal

16. Washington Action Committee

NARUC is very active in developing a variety of educational programs and courses for its commissioner-members. The governmental agencies that regulate utilities and carriers in all 50 states, plus the District of Columbia, Puerto Rico, and the Virgin Islands, are members of NARUC.

PUC organization

State commissions have separate departments for certain functions, such as rates, engineering, accounting, financial, and legal. Usually, these separate departments exist for each type of industry regulated. However, sometimes it is the case that they are combined. For example, electric utility regulation may be combined with the regulation of gas utilities and/or water utilities.

The professional responsibilities of PUC commissioners include:

- Familiarity with the history of utility regulation
- Objectivity
- Familiarity with the industries over which control is exercised
- Discretionary when following procedures (fairness and impartiality in the application of procedures)
- Public information campaigns and education of regulatory control limitations

PUCs are considered to be quasi-judicial and quasi-legislative entities. There is no real separation of power, since they act as administrator, judge, and legislator. Commissions act as administrator when, for instance, they investigate rates or service. In a judicial capacity, PUCs exercise this power when they hold hearings, examine evidence, and make final determinations of a particular utility's conduct.

Commissions can determine the rules they want to enforce at any given point in time. They can then gather evidence against any utility if prosecution is warranted. Finally, when commissions fix rates, for example, they are exercising legislative powers.

Federal Regulatory Entities

Various federal agencies and commissions are charged with the task of regulating the electric utility industry, as well as some other public utilities and other businesses. These federal agencies and commissions have four main duties:

- Formulating a wide variety of relevant rules and regulations
- Ensuring that relevant rules and regulations are implemented
- Investigating and prosecuting cases where there are violations of the relevant rules and regulations
- Presiding over disputes when they arise that pertain to the relevant rules and regulations

When rules and regulations are being implemented, the federal agencies and commissions have the last word. Whatever the agencies and commissions decide is considered to be valid, legal, and binding upon all parties. Federal agencies and commissions will first issue a Notice of Proposed Rulemaking (NOPR) before making a final decision concerning a new or revamped regulation.

Once a NOPR is made, federal agencies and commissions seek industry input in the form of comments. Generally, the governing agency or commission will set a certain time frame in which comments will be accepted. The particular federal body will then review all the comments and will issue its final judgment. Final judgment in cases of new rules and regulations can be challenged by offended parties in the courts or through the filing of a petition to reconsider.

Federal agencies and commissions

At least eight federal regulatory agencies and/or commissions are responsible for regulating the electric utility industry:

1. The Department of Energy (DOE)

2. The Federal Energy Regulatory Commission (FERC)

3. The Environmental Protection Agency (EPA)

4. The Nuclear Regulatory Commission (NRC)

5. The Occupational Safety and Health Administration (OSHA)

6. The Securities and Exchange Commission (SEC)

7. The Department of the Interior (DOI)

8. The Department of Agriculture (DOA)

Department of Energy

This regulatory body directly affects the electric utility industry in that it formulates energy policy and initiatives. DOE has direct control and power over utilities through the EIA, which requires electric utilities and other electricity generators to submit various operation reports. The DOE's policy initiatives are supported in large part by the data that the EIA collects.

The Economic Regulatory Administration (ERA), another division of the DOE, both administers and supervises the DOE's electric utility enforcement activities. The ERA is also charged with overseeing many regulatory programs, granting licenses required for electricity imports from Canada and Mexico, and overseeing the regional coordination of electricity system planning and reliability. In addition, the ERA is the enforcement arm of the Power Plant Industrial Fuel Use Act, a mandate requiring new base load and combined-cycle electricity generating units to be able to burn coal as fuel.

The DOE was created in 1977 with the passage of the Department of Energy Organization Act. The purpose of the act was to consolidate several federal energy agencies under one entity. Many agencies were consolidated under the DOE:

- Energy Research and Development Administration (ERDA)

- Federal Energy Commission (FEC)

- Pittsburgh and Morgantown energy technology centers

- Federal Power Commission (FPC)

- DOI's Bureau of Reclamation power marketing functions

- DOI's former responsibilities for the Alaska, Bonneville, Southeastern, and Southwestern Power Marketing Administrations

- Economic Regulatory Administration

- Energy Information Administration

- Federal Energy Regulatory Commission

The U.S. president appoints a secretary to oversee the many duties and roles of the DOE. The department secretary is the principal adviser to the president on all energy policies.

The DOE, through its secretary, recommends the development of regulatory incentives, regulations, and standards. It also offers recommendations to the president and/or to the Congress on amounts of money needed for specific energy research efforts. The DOE, through all its varied roles and functions, has a tremendous impact over how electric utilities conduct their planning and other business activities.

Federal Energy Regulatory Commission

The FERC is perhaps one of the most influential agencies affecting the operations of electric utilities. The agency is set up as an independent regulatory organization within the DOE. Its decisions regarding the industry are final; the secretary of the DOE has no powers of recall on any of the FERC's decisions. The only action that offended parties may take is to appeal a FERC final ruling to the U.S. Court of Appeals.

Formally created by the Department of Energy Organization Act, FERC's five members are appointed by the U.S. president. The U.S. Senate must confirm the appointments, which run for four years.

FERC has far-reaching regulatory, permitting, and monitoring functions:

- Regulating wholesale electricity sales
- Regulating most merger activities between and among regulated utilities and nonutility companies
- Issuing licenses for nonfederal hydroelectric projects
- Issuing permits for maintenance of facilities located at international borders for electricity transmission between the United States and other countries
- Overseeing certain electric utility stock transactions
- Monitoring relationships electric utility executives have with companies doing business with utilities

Environmental Protection Agency

Much like the DOE, the EPA is an independent cabinet-level government agency. The U.S. president nominates an administrator, a deputy, and nine assistant administrators. The U.S. Senate must confirm the nominations.

Created in 1979, the EPA has as its overall mission the task of protecting and enhancing the environment, both in current times and into the future. All of its policies and decisions reflect this underlying goal.

Within its reach are all environmental factors stemming from pollution, solid waste, pesticides, radiation, and toxic substances. EPA is involved in three main activities:

1. Conducting pertinent research in environmental matters
2. Setting specific environmental standards
3. Monitoring and enforcing all of the environmental standards it promulgates

Nuclear Regulatory Commission

Also an independent agency, the NRC was established in 1975 under the Energy Reorganization Act. Under this Act, the NRC was given many of the duties of its predecessor, the Atomic Energy Commission. One of the NRC's most important duties is that of issuing nuclear power plant permits.

Five commissioners, each appointed by the U.S. president and confirmed by the Senate for five-year terms, administer the NRC. The agency's expansive regulatory function is divided among three operating offices—the Office of Nuclear Reactor Regulation, the Office of Nuclear Material Safety and Safeguards, and the Office of Nuclear Regulatory Research. In addition, there are five regional offices and four independent panels that conduct the business of the NRC.

NRC impacts the electric utility industry in three major areas:

1. Licensing

2. Inspection and enforcement

3. Research and standards

With respect to licensing, the NRC reviews and issues licenses as they relate to the construction and operation of nuclear power plants, as well as other nuclear-related facilities. The NRC is also charged with licensing entities that possess and utilize nuclear materials.

The NRC requires that those activities involving licensed nuclear operations be in compliance with its rules and regulations. This agency can take the required enforcement action against entities and persons found to be in violation of standard rules and procedures.

With its keen eye toward the safety of nuclear operations and materials, the NRC regularly assesses environmental impact. These two very important areas of concern are often the subject of NRC-backed research, as well as programs that become standards in the utility industry.

The Securities and Exchange Commission

The SEC, another independent agency, was created in 1933 to regulate interstate transactions in corporate securities and stock exchanges. The agency is comprised of five members who are nominated by the U.S. president. They serve five-year terms.

The SEC affects electric utilities in some specific ways:

- Requires disclosure of information regarding public offerings of securities.

- Investigates instances (or suspected instances) of securities fraud with the authority to enforce sanctions against violators.

- Used to regulate security sales and purchases, utility properties, and other such assets held by registered public utility holding companies and their respective utility subsidiaries (see next).

- Used to regulate instances where reorganizations, consolidations, and mergers take place within and among utility holding companies (see next).

- Used to require officers, directors, and shareholders who possessed in excess of 10% of a company's registered securities to file reports. These reports had to show the amount of holdings and reflect any alterations made in the amount(s) of ownership (see below).

- Regulates securities trading

As noted previously, the SEC once took an active role in regulating the dealings of certain holding companies, particularly concerning mergers and what types of subsidiary companies holding companies could invest in. The SEC held this power for regulating electric utilities that were defined as registered holding companies under the Public Utility Holding Company Act (PUHCA). Enacted in 1935 as part of the New Deal legislation, PUHCA was repealed on August 8, 2005 as part of a sweeping energy policy reform law known as the Energy Policy Act of 2005 (EPAct of 2005). This comprehensive law is discussed further in chapter 6. However, for purposes of this chapter, it is important to understand how PUHCA has affected certain electric utilities, the electric utility industry as a whole, and consumers for some 70 years.

PUHCA's original intent was to halt corruption and scandals in the electric utility industry during the Great Depression era. It was meant to protect consumers against business dealings that often threatened the reliability of electric utilities. Specifically, PUHCA imposed extensive regulation on the size, spread, business type, and finances of holding companies owning and operating electric utility companies.

Under the now-defunct PUHCA, registered holding companies were subject to SEC monitoring, which included the requirement that the holding company receive SEC approval for all financial transactions among affiliates. The SEC was also responsible for monitoring the capital structures of registered holding companies and their affiliates (as defined in PUHCA). PUHCA also contained mandates allowing for the following:

1. The review of registered holding company boards of directors
2. The limiting of companies' operations to certain geographical boundaries
3. The prohibition of diversification into unrelated lines of businesses

There have been numerous congressional attempts over the years at repealing PUHCA. With the passage of EPAct of 2005, the strongholds of this controversial law have now been loosed. Additional discussion of PUHCA repeal appears later in this chapter and in chapter 6.

The Occupational Safety and Health Administration

OSHA is a federal regulatory agency established by the Occupational Safety and Health Act of 1970 under the Department of Labor. OSHA has broad responsibilities to a variety of different businesses. Its main function is to ensure that some 90 million Americans have safe and healthy conditions in their workplaces. The act was the very first nationwide regulatory program to try to prevent injury and illness on the job.

As originally drafted, the Occupational Health and Safety Act designated that the DOE would have full responsibility for carrying out its own safety and health mandates. However, the DOE turned to OSHA as an independent agent to ensure that a dedicated team of experts would carry out these duties in line with the integrity of the act.

Of particular importance are OSHA's standards relating to electric IOUs and their employees. These standards cover the safe operation of mechanical equipment and hazardous materials, among other things.

The Department of the Interior

Congress created this agency in 1849 for the principal task of monitoring such things as the nation's census count, patent office, and to oversee federal lands. Among the bureaus within DOI that directly affect electric utilities are the Bureau of Land Management (BLM) and the U.S. Fish and Wildlife Service (FWS).

The BLM is also considered the manager of the nation's federal lands. This agency affects electric utilities in that utilities' power lines will often pass through federal lands in order to reach their customers. When this occurs, electric utilities must abide by BLM regulations.

The BLM affects electric utilities more directly in its role as the principal leasing agent for federal lands with vast coal resources.

The FWS protects fish and other wildlife along with their natural habitats. Endangered and threatened species and wildlife management activities conducted by the FWS directly affect electric utilities. As much as possible, electric utilities operate their transmission lines, generation activity, etc. in a manner that impacts wildlife and their natural habitats only minimally.

Also, many endangered or threatened species populate lands where utilities operate. These particular pieces of real estate are managed by the BLM to ensure the wildlife and habitats are properly preserved. It also has been standard practice that electric utilities work with the BLM on such projects as developing guidelines to prevent birds from colliding with power lines.

The Department of Agriculture

The DOA was created in 1862. It includes the U.S. Forest Service, which is a governmental agency charged with managing the national forests and millions of acres of other federal lands. As mentioned earlier, and is the case here, electric utilities' power lines cross these protected federal lands. Electric utilities are required to comply with the regulations set forth by the U.S. Forest Service.

Federal Laws Loosen Their Grip

Multitudes of federal rules, regulations, and laws affect most electric utility companies. These federal laws are in addition to the vast complexity of state-mandated rules and regulations. In most instances, federal laws supersede state laws. In many cases where state law is not as stringent as federal law, it has become a trend over the past decade for electric utilities to adhere to and often exceed that which federal law mandates.

The number and kind of electric utilities range widely in the United States. Thus there are some utilities, such as electric cooperatives and municipal utilities, that generally operate under a different set of regulations than those affecting electric IOUs. It would be impossible to outline all the laws that affect the bulk of the nation's electric utilities. The focus instead concerns how the industry's quest to evolve in the competitive restructuring and reform area has resulted in a pronounced lessening of the tight grip on electric utilities that had previously defined the industry.

PUHCA: stumbling block or safeguard?

The national debate over PUHCA has been ongoing for a number of years. According to the EEI, PUHCA imposes outdated restrictions on the business activities of both electric and gas utility holding companies and has served to act as a barrier to competition. The act also, EEI maintains, deprived consumers of the full range of energy provider services and choices they would have if the act were not on the books.

There are other issues the electric IOU organization points out about PUHCA. The EEI maintains that PUHCA has had the following results:

- Restricted the flow of capital into U.S. energy markets and slowed development of new generation and transmission capacity
- Limited the number of new suppliers in electricity markets by prohibiting exempt wholesale generators from selling directly to retail consumers
- Acted as a barrier to the formation of regional independent transmission companies (ITCs)

According to EEI, FERC is promoting the formation of RTOs as a means for achieving regional electricity markets (see chapter 5). Most market participants, EEI contends, expect that RTOs will be the entities primarily responsible for expanding transmission capacity. "Unfortunately, because PUHCA would apply to ownership or control of multistate regional independent transmission companies, the law poses a barrier and undermines efforts to form them," EEI said in the March 2003 article, *Remove Federal Barriers to Competition: Repeal the Public Utility Holding Company Act.*

However, according to citizen's watchdog group Public Citizen, PUHCA was worth keeping since the act promoted economic growth and stability in the electric utility industry. Without it, the electricity business will become more concentrated, less transparent, less regulated, and more prone to failure, according to Joan Claybrook, president of Public Citizen.

History, in this instance, may repeat itself now that PUHCA has been repealed, Public Citizen maintains. The group points out that prior to PUHCA, there were 53 utility holding company bankruptcies and 23 bank loan defaults. Just since 1992 exemptions from PUHCA, there have been numerous bankruptcies of PUHCA-exempt companies—Mirant, NRG, Enron, NEG—and lots of power plants have been sold at "fire sale" prices.

Prior to repeal of the act, there were more than 100 utility holding companies; however, not all of them were required to be registered under PUHCA. Table 4–1 lists registered holding companies that were under the strictures of PUHCA's provisions.

Companies exempted from PUHCA were still subject to state regulation but were not governed by special SEC requirements, with one exception. The only SEC requirement applying to them was one that stated that even exempt holding companies could not own utility company interests outside their geographic area, or acquire more than 5% of a utility's securities without SEC approval.

Exempt companies were able to enter other lines of business. Registered companies, on the other hand, were prohibited from doing so. Registered companies were strictly confined to their single, integrated business, which may have reached into other businesses. However, those business interests had to be deemed "reasonably incidental" or "economically necessary" as defined by the SEC.

Table 4–1. Holding companies once registered under PUHCA

Registered Company Name	Subsidiary Registered Company (If applies)
AGL Resources Inc.	NUI Corp.
Allegheny Energy Inc.	Allegheny Energy Supply Co. LLC
Alliant Energy Corp.	
Ameren Corp.	
American Electric Power	AEP Utilities Inc.
Black Hills Corp.	
Centerpoint Energy Inc.	Utility Holding LLC
Cinergy Corp.	
Dominion Resources	Consolidated Natural Gas
E. ON ag	E.ON US Holdings GmbH
	E.ON US Investements Corp.
	LG&E Energy LLC
Emera Inc.	Emera U.S. Holdings Inc.
	BHE Holdings Inc.
Energy East Corp.	
Enron Corp.	
Exelon Corp.	Exelon Energy Delivery
	Exelon Energy Ventures Co. LLC
FirstEnergy Corp.	
Great Plains Energy Inc.	
KeySpan Corp.	
National Fuel Gas Co.	
National Grid Transco plc	National Grid (U.S.) Holdings Limited
	National Grid General Partnership
	National Grid USA
	National Grid (U.S.) Investments
	National Grid (U.S.)Partner 1 Limited
	National Grid (U.S.) Partner 2 Limited
	National Grid Holdings Inc.
Nisource Inc.	Columbia Energy Group
Northeast Utilities	
Pepco Holdings Inc.	Connectiv
PNM Resources Inc.	
Progress Energy Inc.	
SCANA Corp.	
Scottish Power plc	Scottish Power NA 1 Limited
	Scottish Power NA 2 Limited
	Scottish NA General Partnership
	PacifiCorp Holdings Inc.
Southern Co.	
Unitil Corp.	
WGL Holdings	
Xcel Energy	

Source: SEC

Further, registered holding companies could not be involved in any way in utility or nonutility businesses, except under certain circumstances. They also were banned from investing in utilities that were not a part of their integrated system. There were other constraints on the few registered public holding companies that existed prior to the repeal of PUHCA. These included provisions that they were prohibited from doing the following, unless SEC approval was sought:

- Participating in any contracts associated with services, sales, or construction when the contracts were with affiliated public utility companies
- Issuing or selling equity securities or altering the priorities, preferences, voting power, or other rights of any existing securities
- Acquiring securities, utility assets, or other interests in any other business
- Borrowing from subsidiary companies or receive credit from any subsidiary public utility company
- Declaring dividends

As might be expected, it is difficult, if not impossible, to know the exact effect the repeal of PUHCA will have on stakeholders in the electric utility industry. This is given the fact that this 70-year-old law, while it affected only a handful of regulated public utility holding companies, had a tremendous effect, as outlined, on the whole of the industry. From electric utilities and the public utility holding companies to consumers and the securities market, this big law with few targets has left a lasting impression. It only remains to be seen how PUHCA's absence will be received and acted upon in the industry.

PURPA: stage set for repeal

The Public Utility Regulatory Policies Act (PURPA) was enacted during the energy crisis of the 1970s. Congress reacted to this crisis by taking steps to reduce the nation's dependence on foreign oil and on fossil fuels in general. Congress' goal was also to diversify the technologies used for electricity generation by cultivating conservation and the efficient use of resources. With the signing of EPAct of 2005, conditions have been set into motion for PURPA's repeal.

By most standards, PURPA has been successful in its mission. The act promoted cogeneration, as well as the use of renewable resources and other energy-efficient technologies. PURPA also introduced competition by demonstrating that electricity generation was not a natural monopoly.

Facilities meeting PURPA's requirements are called *qualifying facilities* (QFs). The act encourages QF construction through the requirement that electric utilities purchase QF electric output at a price no greater than the cost the utility would have incurred had it supplied the power itself or obtained it from another source. EPAct of 2005 established conditions for eliminating PURPA's mandatory purchase obligation and revising the criteria for new QFs that seek to sell power under the mandatory purchase obligation.

PURPA, like PUHCA, has been targeted for repeal because of the electric utility industry's move toward reregulation and competition. Is it a needed form of regulation? As is the case with PUHCA, PURPA has undergone some scrutiny as well as praise in the industry. Proponents of PURPA claim that there is no guarantee a free market can sustain the goals of PURPA, especially in the use of cogeneration and renewable energy sources. There are other arguments in favor of retaining PURPA:

- The nation must be able to handle another energy crisis through fuel diversity.

- Incentives must remain in place to conserve energy and to use more environmentally friendly fuels.

- QFs bring increased reliability and decrease the need for large costly plants.

- At this point, utilities still have too much market power, and PURPA levels the playing field for nonutilities.

- Immediate repeal is a piecemeal approach; repeal should be included in comprehensive industry restructuring legislation.

Those in favor of PURPA repeal claim that the act is anticompetitive because utilities are required to purchase from QFs. In addition, repeal proponents claim that the 1992 Energy Policy Act (discussed next in this chapter) has provisions for exempt wholesale generators that render PURPA obsolete.

Other reasons are cited for repeal:

- PURPA has resulted in high prices to consumers because QF contract terms were lengthy and were based on erroneous forecasts of high capital costs and increases in demand and the price of natural gas.
- PURPA's goals have already been achieved.
- Cogenerators and renewable energy sources have already gotten a foothold and do not need further promotion.
- Immediate repeal is necessary; it will take too long if it is contained in comprehensive industry restructuring legislation.

Energy Policy Act of 1992 begins new era

The national Energy Policy Act of 1992 (EPAct of 1992) is viewed by many as a turning point in the evolution of the electric utility industry. Significant alterations were made to PUHCA with its passage. The most important change was a loosening of the tight grip PUHCA had on registered public holding companies. EPAct of 1992 allowed registered public utility holding companies to diversify into such areas as independent power production and others.

Perhaps of most significance is the fact that EPAct of 1992 set the stage for competition in the electric utility industry. Its thrust was to promote economic development and energy conservation, but it had a major goal of encouraging competition. EPAct began what has become a tidal wave of changes in national policies and regulations that have, in the past, defined the place of utilities in the generation, transmission, and sale of electricity.

EPAct of 1992 revised PUHCA to include a new class of power producer, called *exempt wholesale generators* (EWGs). This new class of producer could be owned by a utility, a utility holding company, or a developer not necessarily connected to a utility.

Under EPAct of 1992, the generation market is substantially freed up. EWGs are able to compete with experienced utilities that have been in the electric generation business for years. Furthermore, the act permits registered electric utility holding companies to operate EWGs anywhere in the United States and to reach out into foreign markets to purchase or develop generating facilities.

As can be seen from the above discussion, EPAct of 2005 is picking up where EPAct of 1992 left off—changing and altering much of what the earlier EPAct had set into motion.

Wholesale and retail wheeling

One of the biggest challenges that the 1992 act presented was that it required utilities to allow wholesale electricity generators to access their power lines. This is termed *wholesale wheeling*. It is encouraged under the provisions of the act and is enforced by FERC. FERC has the power to order utilities to comply with the provisions of EPAct if it determines that certain conditions have been met.

Wheeling is the transmission of electricity from a seller to buyer over the transmission lines of one or more other entities, including utilities and government facilities. Wholesale wheeling occurs when the buyer of the power resells the wheeled power to retail customers. The practice of wholesale wheeling has been in existence since the passage of the Federal Power Act, which was enacted in 1920. However, it was not until EPAct of 1992 was passed that wholesale wheeling opened up new opportunities. These opportunities allowed utilities to wheel power among themselves and opened the door for nontraditional players to try their hand at the practice. It is important to take note that while the new EPAct of 2005 is loosening strictures and actually encouraging the electric utility industry to continue in its restructuring efforts, this latest act in no way will impact much of what the earlier act started. The 2005 act is, in essence, an extension and an enabling piece in the restructuring pie.

The following chapters will explain how reregulation and competition have occurred in the utility market and how they will likely progress in the future. The only thing about the utility industry that does remain constant is change, and with the next few chapters, this will be examined fully.

Restructuring, Standards, and Accountability

5

The very nature of the electric power industry is one of stability and of on-demand performance. Electricity consumers, whether residential, wholesale, or retail, expect nothing less of this once-staid industry.

Thought of for years as one of the most powerful monopolistic enterprises, the electric utility industry since the early 1990s has been undergoing significant, albeit gradual, structural change. Ultimately, the changes under way will lead this iconic industry down a path towards an end that cannot yet be seen. Among a plethora of concerns in this new electric marketplace, the general and pertinent issues of stability and performance remain at the top of the "to accomplish" list.

At issue in the restructuring of the electric power industry is an increase in competition in the generation and retail sales components. Structurally, this reorganization will change the way electricity is priced, traded, and marketed in the United States.

To further explain this ongoing change, it is important to understand that the electric utility industry in its traditional form was operated as a vertically integrated structure of generators, transmission lines, and distribution facilities. The transformation, already in motion, is separating generation, transmission, and distribution into distinct and separate entities. Under this reorganized operation, generators compete for sales across common transmission lines to local distribution outlets.

As one could imagine, the stakes are high in this game of change. As will be presented in this and other chapters ahead, it will become clear that there are many issues to address as restructuring continues to take hold of the utility industry. One of the most glaring up-front ramifications is how generators are in direct competition with each other to serve more than one utility distributor. Generators stepping up to the plate with the least-cost power will win the game of serving more utility customers. There is no guarantee that all generators will have the same distributor next year or the year after. The only constant is that generators' power distributors will likely change from year to year. The real grand prize winners, proponents of restructuring claim, are utility customers, since ideally they will reap the benefits of lower electricity bills.

The Driving Forces Behind Competition

To say that competition is happening in the electric utility industry for any single reason would be misleading. Economic as well as technological factors are at play in the push for change within the industry. There are three main forces at work:

1. A general reevaluation of regulated industries and a rethinking of how the introduction of competition might improve efficiencies

2. The wide disparity of electricity rates across the United States

3. Technological improvements in gas turbines, which have changed the economics of power production

As shown in figure 5–1, there are three major reregulation policy drivers. These include the EPAct of 1992, FERC Order 888 and Order 889 of 1996, and finally, the EPAct of 2005. Chapter 6 discusses the key provisions of EPAct of 2005.

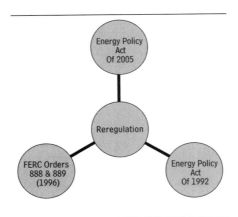

Fig. 5–1. Reregulation policy drivers

The Role of FERC

The FERC's adoption and implementation of EPAct of 1992 and its adoption of retail access plans by a growing number of states (discussed below) got the proverbial ball rolling. It introduced greater competition in the generation and retail supply segments of the utility industry. Table 5–1 lists FERC's major responsibilities.

Table 5–1. FERC's role in the electric utility industry

	Approve	Review	Certify
Rates for wholesale sales of electricity and transmission in interstate commerce for jurisdictional utilities, power marketers, power pools, power exchanges, and Independent System Operators (ISOs)	X		
Issuance of certain stock and debt securities, assumption of obligations and liabilities, and mergers		X	
Officer and director positions held between top officials at utility companies and certain firms with which they do business		X	
Rates as set by the federal power marketing administrations		X	
Exempt Wholesale Generator status		X	
Qualifying small power production and cogeneration facilities			X

EPAct of 1992 amended the 1920 Federal Power Act to authorize the FERC to order public utilities to provide transmission services for competitive wholesale power purchases and sales. Prior to EPAct of 1992, the FERC could not mandate an electric utility to provide wheeling services for wholesale electric trade.

By way of explanation, *wheeling* is the transmission of electricity for a third party. EPAct of 1992 required electric utilities to allow outside wholesale generators to have access to their power lines. The FERC can order utilities to do so if it decides that certain conditions have been met. Ultimately, this meant that generators could make sales for resale to noncontiguous utilities.

In 1996, the FERC issued the historic FERC Order 888 and Order 889. In short, these orders were issued to prevent undue discrimination in the provision of transmission services. These orders opened the floodgates to the industry's most massive reform ever undertaken.

Order 888 guaranteed suppliers and wholesale purchasers access to transmission-owning utilities. This order also provided for utility recovery of costs that might be stranded as a result of open access. Potentially *stranded costs* are costs that utilities would have had the opportunity to recover at expected market prices.

As a result of industry concerns over how the FERC should deal with the transition costs accompanying the shift to competition, FERC issued Order 888-A in early 1997. This order's goal was to achieve a balance between the different approaches on how to achieve the recovery of stranded costs. Also addressed was the critical need of maintaining the financial health of the industry, maintaining the regulatory deals of large past investments, and avoiding shifting the costs to customers.

Order 889 required public utilities owning or operating transmission facilities to establish electronic information systems, known as *open access same-time information systems* (OASIS). The practical outcome of establishing OASIS was to provide all parties with identical access to information on available transmission capacity. In addition, Order 889 required electric utilities to implement standards of conduct that functionally separated the operation of the transmission system from each utility's wholesale merchant function.

FERC Order 888's open access provisions have reduced barriers to FERC approval of market-based rates for wholesale power sales. Since the FERC began approving market-based pricing in 1988, the major impediment has been the potential for utilities to exercise market power through ownership or control of transmission facilities.

It stands to reason that the filing of an Order 888 open access transmission tariff (discussed below) meets FERC's standards with respect to mitigating market power in transmission. With this barrier removed, the FERC has approved market-based rates for more than 300 electric utilities and power marketers.

North American Electric Reliability Council

The issue of reliability has also been a major, if not overriding, concern in the move to keep power flowing despite the many structural changes within the industry. NERC has this task well at hand. Its mission is to ensure that the bulk electric power system in North America is reliable, adequate, and secure.

To this end, NERC's board of trustees in February 2005 unanimously agreed to adopt a comprehensive set of reliability standards for the bulk electric system. The new reliability standards were presented in 14 categories (as shown in table 5–2) that set forth 91 new standards. The detailed standards can be found by visiting the following Web site: www.nerc.com/~filez/standards/Reliability_Standards.html.

Table 5–2. 2005 comprehensive reliability standards categories

Resource and Demand Balancing	Modeling, Data, and Analysis
Critical Infrastructure Protection	Organization Certification
Communications	Personnel Performance, Training, and
Emergency Preparedness and Operations	Qualifications
Facilities Design, Connections and Maintenance	Protection and Control
Interchange Scheduling and Coordination	Transmission Operations
Interconnection Reliability	Transmission Planning
Operations and Coordination	Voltage and Reactive

Source: NERC

The reliability standards, which went into effect April 1, 2005, incorporated the existing NERC operating policies, planning standards, and compliance requirements into an integrated and comprehensive set of measurable reliability standards. These standards apply to all entities that play a role in maintaining the reliability of the bulk electric system in the United States and in Canada.

NERC's efforts at formulating, adopting, and implementing a comprehensive set of reliability standards came in the wake of the August 14, 2003 electric power blackout. This blackout cascaded across large portions of the Northeast and Midwest United States and Ontario, Canada. Termed one of the worst in recorded history, it affected an estimated 50 million people and disrupted 61,800 MW of electric supply. More information on this blackout can be found in chapter 7.

On April 5, 2004, a joint U.S.-Canadian task force issued a final report on its blackout investigation and identified the causes of the blackout. The blackout, the report stated, stemmed from the fact that several entities had violated NERC operating policies and planning standards. There were four general groups of causes cited as producing the blackout:

1. Inadequate system understanding
2. Inadequate situational awareness
3. Inadequate tree trimming
4. Inadequate reliability coordinator diagnostic support

Further, the report found that due to a variety of institutional issues, the NERC standards were unclear, ambiguous, and nonspecific, making it possible for bulk power system participants to interpret the standards in varying ways.

NERC's 2005 comprehensive reliability standards are aimed at addressing the task force's findings. NERC has also established a Reliability Division within its Office of Markets, Tariffs, and Rates.

In a February 9, 2005 press release, FERC said, "In sum, the Commission expects public utilities to comply with NERC reliability standards and to remedy any deficiencies identified in NERC compliance audit reports and recommendations. The Commission will consider taking utility-specific action on a case-by-case basis to address significant reliability problems or compliance with Good Utility Practices, consistent with its authority. A failure to comply with such industry standards could in some circumstances affect Commission determinations as to whether rates are just and reasonable. For example, it may be appropriate to deny full cost recovery in circumstances where a transmission provider fails to provide full reliability of service."

The Advent of the ISO and RTO

Both power pools and groups of utilities in most regions of the United States have responded to FERC and NERC rulemakings by forming or proposing to form ISOs or RTOs, as shown in figure 5–2. The FERC, in particular, has been promoting the voluntary formation of RTOs and ISOs with the ultimate goal of eliminating discrimination in access to the electric grid.

The main goal of these groups, although ISOs and RTOs have more specific goals as outlined, is to ensure nondiscriminatory operation of transmission systems and to facilitate the development of regional transmission tariffs. (As with FERC and the NERC, as discussed later, ISOs and RTOs are self-funding—participants pay for their involvement and support of the organization's efforts.) FERC Order 888 mandated this, and it is referred to as *comparable service*. That order, as discussed earlier, requires utilities owning bulk power transmission facilities to treat any of their own new wholesale sales and purchases of energy over their own transmission facilities to the same transmission tariffs they apply to others.

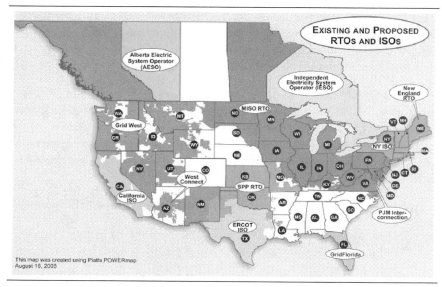

Fig. 5–2. Existing and proposed RTOs and ISOs

The EIA claims that advantages are expected to arise from the operational efficiencies resulting from the oversight of a large regional transmission system and from the elimination of multiple tariffs. However, as EIA points out, this program is not without detractors who claim that advantages may still go to vertically integrated utilities that maintain transmission ownership rights as opposed to nonowners. One possible effect is that ISOs will curtail needed future transmission facility expansion.

According to NERC, ISOs are being formed to facilitate open access of the electric transmission system, manage dispatch of generation and congestion, and administer a regional spot market for energy, capacity, and ancillary services. The coordination of these three activities under a centralized operation serves to enhance the stability and reliability of the regional power grid. The FERC developed both the ISO and the RTO concepts to help support open transmission access and wholesale competition within the industry.

An RTO is comparable to an ISO in that it facilitates electricity transmission on a regional basis, manages generation dispatch and congestion, and administers organized day-ahead and real-time spot markets. An RTO is designed to create efficiencies in the competitive wholesale market by accurately determining the number of paths through which electricity must flow. As discussed earlier, the principal intent of both ISOs and RTOs is to reduce the pancaking or cumulative cost additions of different rate structures, thus reducing the overall cost of electricity transmission.

The FERC approved the Midwest Independent Transmission System Operator Inc. (MISO) as the nation's first RTO in December 2001. In July 2005, MISO announced its joint operating agreement with the PJM Interconnection to move forward on the implementation of a nondiscriminatory joint and common electricity market (JCM) covering the two RTOs' collective regions.

According to MISO–PJM, there are benefits from their joint collaboration:

- **Decreased production costs.** It allows reduced overall production costs across the combined footprint as a result of increased ability to do business between the markets.

- **Decreased staffing costs.** It allows elimination of efforts associated with performing some tasks.

- **Decreased computing maintenance costs.** It allows an elimination of hardware, software, and networking devices/ connectivity.

- **Decreased facility costs.** It allows elimination of office space, equipment, and/or facility.

- **Risk avoidance.** It eliminates risks associated with poor performance of market functionality or uncertainty associated with market information.

- **Consistency.** It provides consistent results for market participants' activities across MISO and PJM.

- **Opportunity.** It enhances ability to interact with the markets and to increase shareholder value, or improve the bottom line of participant organizations.

States Respond to Changes

Restructuring at both federal and state levels has transformed the generation and retail supply segments of the electric power industry. Competitive markets will increasingly replace state and federal regulators in setting the price and terms of electric generation and supply services.

State legislatures and/or PUCs in most states are considering or have approved plans allowing retail customers direct access to generation markets by allowing customers to choose among competitive suppliers of generation. Some regions may establish generation tracking and disclosure systems, providing consumers with the option of purchasing from suppliers of renewable or other preferred types of generation.

Furthermore, a number of states have adopted legislation that will make retail access available to their customers. Pilot programs to initiate and evaluate retail access are being conducted in many states where retail access plans are approved or are likely to be approved in the near future.

In some states, retail access plans face legal challenges related to the recovery of potentially stranded costs and other issues. *Stranded costs* are those costs that refer to a utility's fixed costs, usually related to investments in their generation facilities. These fixed costs would no longer be paid by customers through the rates they pay in the event that customers opt to purchase power from other suppliers.

From the foregoing discussion, it is clear that restructuring is inextricably linked to standards of performance and operation as well as to accountability. FERC, as explained, has regulatory venue of jurisdiction over electricity sales, wholesale electric rates, hydroelectric licensing, and other forms of jurisdiction. NERC's aim is the promotion of reliability and adequacy of the bulk power supply in the electric utility systems of North America. These are not competing forces in the electric utility industry of today and the future, but rather are complimentary entities. Next, the discussion turns to the adoption of a landmark piece of legislation, which in some cases elicits more (and in some cases commands less) accountability from the nation's electric utilities.

Part III:
Emerging Issues
and Trends

The Energy Policy Act of 2005

It makes sense that as large and critical as the electric utility industry is in the United States, federal or state laws and regulations governing its operations are not put into motion often. For this reason, this chapter will be devoted to discussion and potential ramifications of the Energy Policy Act of 2005 (EPAct of 2005).

Few could rebut the fact that EPAct of 2005 is one of the most important, far-reaching forms of law enacted since the passage of the Energy Policy Act of 1992. EPAct of 2005 was not perfect and did not address every single concern of those in the industry supporting the various reforms contained within it. However, EPAct of 2005 carries with it sweeping industry implications and sets the tone for a much-changed energy strategy for the 21st century.

Legislative Overview

EPAct of 2005 stemmed from HR 6, a 1,724-page piece of legislation that Congress had been hammering on since 2001. The U.S. House of Representatives passed HR 6 by a vote of 275 to 156 on July 28, 2005, while the U.S. Senate cleared the piece of ground-breaking legislation on July 29, 2005 by a vote of 74 to 26. The president of the United States signed the bill into law August 8, 2005.

The signing of HR 6 into law will be costly—some $12.3 billion over 10 years. However, most in the industry agree that its provisions are sorely needed in an industry where massive restructuring has left gaping holes in current laws addressing new issues and trends emerging from the reregulation of the power industry. However vital the legislation, it remains unclear how long it will take for many of its provisions to come to fruition.

The bill was touted by Sen. Pete Domenici (R), chair of the Senate Energy and Natural Resources Committee, as not a bill for today or tomorrow, but a "bill for the future." Similarly, the president of the United States said the bill would not solve the nation's energy problems overnight, but that the new law would help to begin now in addressing those problems.

Purpose of bill hinges on reliability

Broadly stated, the purpose of EPAct of 2005 is to encourage much-needed investment in the nation's energy infrastructure while placing an emphasis on measures that will stimulate energy efficiency and conservation. The law will help to increase electric reliability while promoting a diverse mix of fuels to generate electricity, and it aims to beef up protections for electricity consumers.

The reliability of the U.S. electricity system was a central, recurring theme during the four years EPAct was under review. This is perhaps at the top of the list as to why this groundbreaking legislation came into being.

The new law will aim to accomplish this task through making electric reliability rules mandatory on all market participants. A self-regulating reliability organization, with FERC oversight, will be created to enforce reliability rules.

Further, the bill grants FERC limited backstop authority to site electric transmission facilities located in national interest electric transmission corridors if states cannot or will not act. The DOE will be given the task of coordinating the federal permitting process for transmission facilities.

Another measure in this area is that the bill requires federal land agencies to expedite approvals for vegetation management activities on rights-of-way (ROWs) to meet mandatory reliability standards.

PUHCA repeal and PURPA reform

Another reason for the comprehensive energy bill was the realization that significant investment was needed in the U.S. energy infrastructure system. To this end, PUHCA, as discussed briefly in chapter 4, was repealed. More details on the historic repeal of PUHCA will be discussed at the end of this chapter. With PUHCA's repeal, certain consumer protections will revert to the authority of FERC and the states. As can be imagined, this will open up a realm of responsibility and accountability that has never before been encountered in the electric utility industry.

Also noteworthy is that the energy bill has set conditions for eliminating PURPA. As was discussed in chapter 4, PURPA was enacted in 1978. The goal was to promote energy efficiency and increased use of alternative energy sources by encouraging companies to build cogeneration facilities and renewable energy projects. PURPA refers to those facilities that meet its requirements as QFs. PURPA encourages their construction by requiring utilities to purchase QFs' electric output at a price no greater than the cost that utilities would have incurred had they supplied the power themselves or obtained it from another source.

The energy bill has set conditions in motion for eliminating PURPA's mandatory purchase obligation, and revises the criteria for new QFs that seek to sell power under the mandatory purchase obligation. More discussion on PURPA reform appears later in this chapter.

Transmission and environmental incentives

Also in the area of energy infrastructure system investment incentives, the bill requires FERC to provide transmission investment incentives and to assure cost recovery for reliability investments that utilities make. The bill also reduces the depreciable lives for electric transmission lines (at 69 kV or higher voltage levels) and natural gas distribution pipelines from the current 20 years to 15 years.

The bill also reduces the cost recovery period from 20 years to 7 years for pollution control equipment on power plants and establishes a new production tax credit for electricity produced at new nuclear energy facilities (see fig. 6–1).

Another infrastructure support tax provision would allow companies that sell transmission assets to a FERC-approved transmission organization to defer the gain on the sale of those assets by one extra year.

Fig. 6–1. New tax credit in store for electricity produced at new nuclear energy facilities

Electricity supply diversity

Fuel diversity in the electric utility industry sector, particularly for the future, has become an important issue for electric utilities. It is with diversity in fuel sources that U.S. electricity providers can provide for a more stable, reliable flow of electricity to consumers.

In addressing this ongoing concern for stability, the bill authorizes federal funding for CCP initiatives and coal-based gasification technologies. These technologies, as discussed in chapter 1, provide the means for electric utilities to utilize the United States' most abundant energy resources.

There are other aspects to the bill:

- Provision for new investment tax credits for clean coal facilities producing electricity, including a tax credit for IGCC projects.

- Improvement and modernization of the tax treatment of the trust funds set aside for nuclear power plant decommissioning.

- Improvement of the unreasonable mandatory licensing conditions that place hydropower, a vital renewable energy source, at risk, while also preserving existing environmental protection standards.

- Extension of the placed-in-service date for renewable energy resource through December 31, 2007 for QFs (wind, closed-loop biomass facilities, geothermal, as shown in fig. 6–2, small irrigation power, landfill gas, and trash combustion facilities).

- Provision of incentives to increase production from areas already open to oil and natural gas production.

- Establishment of a climate change technology program directing the secretary of energy to lead an interagency process to develop and implement a national climate technology strategy. Using greenhouse gas intensity as a measure of success, the energy bill creates incentives for innovative technologies and encourages partnerships with developing nations.

Fig. 6–2. Renewable energy resources, such as geothermal, are given special attention in the energy bill.

General consumer protections

The bill seeks to provide a higher level of protection to electricity consumers and the markets. This is accomplished by providing FERC with greater enforcement authority and the ability to impose higher penalties on companies that violate the FPA (see chapter 4).

There are other general consumer protections included in the bill:

- The FTC is given authority to issue rules protecting consumer privacy and prohibiting unfair trade practices.
- FERC is provided with broad authority to prevent market manipulation.
- FERC is given more explicit direction on preventing harmful cross-subsidization when reviewing electric utility mergers.

Power Delivery

Electric power consumers are given much consideration in the numerous provisions of EPAct of 2005. Restructuring activities, including policies from federal to state levels, have afforded U.S. electricity consumers with an expanded involvement in the manner in which power is delivered to them. Nearly all the provisions aimed at electricity consumers are in the name of choice and cost-saving initiatives.

EPAct of 2005 directs electric utilities to offer their customer classes (commercial, industrial, and residential) time-based rates and metering, methods for demand response, and improved interconnection for power generation assets. These three provisions together are expected to have profound effects on the evolving electric utility industry.

First, time-based rates, along with the metering technology to support them, offer consumers the option to change their electricity usage patterns to off-peak time periods. The purpose is cost savings. State regulatory entities will have the option of adopting minimum time-based rate standards. They are expected to write a report by mid-2007 outlining their standards or, in the alternative, explaining why they chose not to adopt minimum standards.

Time-based options for utilities include offering seasonal rates (varying prices for winter, spring, summer, and fall) and real-time pricing (multiple rates throughout the course of one day). Another option is to offer different rates on different days depending upon forecasted demand and load curtailment (consumers restrict their usage during certain periods and receive credits or lower rates).

EPAct of 2005 requires electric utilities to offer time-based meters to any customer who requests the use of time-based rates. Today's technology affords electric utilities the option of retrofitting many of their current meters to be read automatically. As well, technology has enhanced billing software such that electric utilities may more easily track and manage a plethora of time-of-use rate structures.

Demand response will also become a trend for the future, under the provisions of the new act. This action requires utility customers to limit or completely halt any energy demand during peak load periods. Utility customers and their utilities would have an agreement that would be binding upon customers. There is a direct benefit of demand response—it actually improves the reliability of the overall distribution system and, in some cases, will defer the need for utilities to build generation facilities that are only put into service during high-demand times. State regulators must present a plan by 2007, either explaining their state standard or why they chose not to implement a state standard.

Since nationwide electric utility industry restructuring efforts are geared at revamping the generation side of the business, it makes sense that the act requires state regulators and electric utilities to examine how they might improve power generation. Some of the requirements included are given here:

1. Developing and implementing a plan to increase the efficiency of fossil fuel generation.

2. Developing a fuel diversity plan, wherein electric utilities must discover ways to reduce dependence upon one major fuel source for power generation.

3. Providing a way for distributed generation customers (those who generate their own power on-site) to connect their power generation to the distribution grid. This proposes to be a win–win situation for electric utilities and their customers. EPAct eases the requirement that utilities purchase power created by cogeneration facilities or small power producers (sources of distributed power). That rule, in some cases, caused electric utilities to pay above-market prices for power. Now state PUCs can determine if a distributed power generator customer has access to the distribution, transmission, and energy trading markets. If so, then that customer can sell power at existing market rates through existing market structures, thereby producing benefits for both utilities and consumers.

Provisions Specific
to Rural Electric Cooperatives

Rural electric cooperatives were addressed separately from the bulk of the nation's electric utilities. Co-ops are consumer-owned electric utilities and were established under the Rural Electrification Act of 1936 to extend electric service to rural communities and farms where it is generally more expensive to provide service.

The energy bill broadly relaxes many of FERC's new regulatory burdens for co-ops and will also serve to reduce regulation for some co-ops. Specifically, the bill contains language that codifies case law that establishes the current exemption from most FERC regulation for co-ops owning transmission that are Rural Utilities Service (RUS) borrowers.

All small electric co-ops that sell less than 4 million MWh of electricity a year will be exempt from most FERC regulation, the bill's language stipulates. It exempts co-ops from a new provision granting FERC limited authority to regulate short-term wholesale sales by large nonjurisdictional utilities.

Additionally, the energy bill grants FERC limited authority over larger transmission-owning co-ops with RUS loans to provide open-access transmission service (FERC-lite). However, it exempts small co-ops that sell less than 4 million MWh of electricity per year. Previously, FERC impacted co-ops' transmission activities through existing statutory authority. Ultimately, this process established a case-by-case complaint process and applied existing regulations that required co-ops to provide access to their transmission facilities based on reciprocity. FERC-lite, under the energy bill, establishes a third way for FERC to impact co-op transmission activities.

The energy bill makes provision for FERC to assess the effects of these exemptions for co-ops on consumers and the reliability of electric transmission networks. FERC will be required to conduct the assessment every five years.

The bill also acts to protect the reliability of the bulk transmission system, as referred to above, and minimizes other new regulatory burdens on co-ops. Under EPAct of 2005, there is not a net metering, renewable portfolio standard, nor are there environmental mandates or other major new regulatory programs for co-ops (and other utilities) to address.

Another section of EPAct that favors electric co-ops is the creation of a single national industry-based reliability organization with the authority (subject to FERC oversight) to establish and enforce national reliability standards. There is a $50 million cap on the amount it can raise through fees, dues, and other charges.

The National Rural Electric Cooperative Association (NRECA) is a national association representing the nation's electric co-ops. It also has found favor for its members in the "native load" protection provisions intended to ensure that consumers who have paid for transmission continue to have access to that transmission. EPAct includes language NRECA sought that has several benefits:

1. Co-ops are ensured the same benefits as electric IOUs.

2. Transmission-dependent utilities are ensured the same benefits as transmission owners.

3. Co-ops are ensured the ability to complain to FERC if public utilities hoard transmission under the guise of reserving transmission for native load.

Tax Implications

EPAct of 2005 contains a wide variety of tax provisions, which are intended to aid energy exploration, delivery, and conservation in the electric utility industry. One such provision offers tax credits for the purchase of hybrid, fuel cell, advanced lean burn diesel, and other alternative power vehicles. This credit will apply to vehicles placed in service after 2005, with termination dates varying with the type of alternative power vehicle used.

For those consumers wanting to purchase gas-electric vehicles, or so-called *hybrid* vehicles, the energy bill may make this option more attractive. Hybrids combine an electric motor with an internal-combustion engine to save fuel. It has been estimated that consumers under the hybrid tax credit contained in the new law could cut their tax bills by $1,700 to $3,000, depending upon the vehicle model they purchase beginning January 1, 2006. This particular tax credit will also be extended to the new generation of fuel-saving diesel cars, which are expected to hit the roads in the near future.

Further, the new energy law offers a 30% tax credit for the purchase of qualifying residential solar water heating, photovoltaic equipment, and fuel cell property. The law sets the maximum credit at $2,000 for solar equipment and at $500 for each kilowatt of capacity for fuel cells. This credit applies to property placed in service after 2005 but before 2008.

Business and personal tax credits are also addressed. There will be a new 30% business tax credit for the purchase of fuel cell power plants and a 10% credit for the purchase of stationary microturbine power plants. These credits will be effective for periods after December 31, 2005 and before January 1, 2008 for property placed into service in tax years after December 31, 2005. The new 10% personal tax credit for energy-efficient improvements to existing homes will benefit some consumers. Under this provision, the lifetime maximum credit per taxpayer will be $500 and applies to property placed into service after December 31, 2005 and before January 1, 2008.

Another business tax credit comes in the form of a credit for the construction of new energy-efficient homes. The credit is either $2,000 or $1,000 per home, depending upon the type of home and the energy reduction standard it meets. This credit will apply to homes whose construction is substantially completed after December 31, 2005, purchased by a consumer after December 31, 2005 and before January 1, 2008.

The new energy law also codifies a new deduction for energy-efficient commercial buildings that meet a 50% energy reduction standard. The deduction, which will generally be $1.80 per square foot, and in some instances less than that, is effective for property placed into service after December 31, 2005 and before January 1, 2008. And, finally, energy-efficient dishwashers, clothes washers, and refrigerators are also given special mention. New manufacturers' tax credits will be available at varying amounts on these and other appliances that are manufactured in 2006 and 2007.

Energy Production Tax Credits

The energy bill includes several incentives for energy production, including a 15-year write-off for natural gas distribution lines as well as a 15-year write-off for certain assets used in the transmission of electricity for sale and related land improvements.

Also of interest is a new production tax credit for qualifying advanced nuclear power facilities, and an elective five-year carryback of net operating losses for certain electric companies of up to 20% of the cost of electric transmission capital and pollution control expenses.

Another such provision extends the renewable electricity production tax credit through December 31, 2008 for wind, closed-loop biomass, open-loop biomass, geothermal, small irrigation power, landfill gas, and trash combustion facilities. It is important to note that the energy bill did not extend the terminating placed-in-service date for solar facilities or refined coal facilities. Those termination dates will remain in 2005 and 2008, respectively.

Other major production incentives include the addition of two new qualifying energy sources—hydropower and coal from qualified Indian coal facilities—and creating parity in the credit duration (10 years) for all qualifying sources of energy. This credit has also been extended to incremental hydropower and has introduced a credit passthrough for electric co-op members.

There are other production tax incentive highlights:

- Tax credits for investments in clean coal facilities—20% for industrial gasification or IGCC and 15% for other advanced coal-based projects.

- For air pollution control facilities placed in service after April 11, 2005, EPAct has extended an 84-month amortization for the cost of power-plant air pollution control equipment.

- Inclusion of the tax credit for fuel produced from nonconventional sources in the general business credit. This will result in 1-year carryback, 20-year carry forward for unused credits for tax years ending after December 31, 2005. This tax credit is also extended to coke and coke gas from qualified facilities placed in service before January 1, 1993 or after June 30, 1998 and before January 1, 2010. (The credit applies to production during a 4-year period beginning on the later of January 1, 2006, or the date the facility is placed in service.)

PUHCA Repeal:
70 Years of Structure Comes to an End

EPAct of 2005 is historic. One of the biggest reasons is small in wordage in the act but huge on potential impact. These words are "The Public Utility Holding Company Act of 1935 (15 U.S.C. 79 et seq.) is repealed." The potential impact is much like the spokes on a wagon wheel—tangents that range from an increased climate of M&A to the likely scenario of stepped-up investments in power transmission and production. On the downside, the possibility exists that repeal may open up potential corruption and resistance to regulation.

Proponents of PUHCA repeal claimed that the dated regulations were hampering the ability of electric utilities to provide the benefits that could arise from significant investments and economies of scale. Effectively, with the repeal of PUHCA, what this means is now companies that are comprised of dozens or hundreds of small, local utilities will have the resources required to build new power lines and generation facilities. Prior to PUHCA repeal, this would simply not have been possible.

PUHCA repeal opponents had several concerns. The most voiced concern was that absent the regulations provided by the 1935 act, the doors would be wide open for mass consolidation and buyout by multinational companies. Opponents claimed that these possibilities may make it even more difficult for both state and federal regulators to rein in potential "unruly" utilities.

Historical recap

To recap from chapter 4, PUHCA was part of the New Deal legislation passed in 1935 in response to corruption and scandals in the energy companies of the time. Its intent was to protect consumers from business deals that could threaten the reliability of electric utilities. After 70 years, these protections no longer exist.

Specifically, PUHCA brought about extensive regulation of the size, spread, business type, and finances of the holding companies that owned and operated electric utilities.

Focus on M&A. To look at the language of PUHCA, it becomes readily apparent that its main focus was on the activities surrounding and resulting from the combination of two or more electric utilities. Any company wishing to merge or acquire other companies was subject to rules on the location of the merging companies, the diversity of holdings the postmerger entity could have, and the amount of debt the resulting company could hold. The SEC scrutinized and gave an up-or-down vote on all proposed mergers and acquisitions.

Risk minimized. PUHCA's strictures made it difficult for energy holding companies to have any involvement in risky businesses. As part of this rule, any companies that sought to become owners of public utilities had to divest themselves of their nonutility holdings. Proponents of PUHCA claim that as a result of this requirement, not one PUHCA-regulated utility holding company has gone bankrupt since that law's adoption in 1935.

Likely repercussions

While it would be difficult, if not impossible, to know every single impact the repeal of PUHCA will hold for the electric utility industry, it is worth at least some mention of speculation.

Perhaps the greatest effect of repeal is that the electric utility market will be opened up for new ownership potential. EPAct of 2005 has removed the age-old barrier to long-distance business combinations between utilities. It is expected that most of the near-term merger activity will be among the small- to medium-sized utilities. This is due in part to the fact that they will have the greatest number of potential merger candidates in the United States. For larger utilities, it has been postulated that opportunities will be opened up for them, potentially in the longer term, to consider investments in more distant utility assets as well as companies within the United States.

On the flip side, foreign companies—both utilities and nonutilities—will now have the opportunity to acquire electric and gas distribution companies in the United States. They will be able to do so without the fear of extensive additional regulation that for so long has acted as a major hurdle for their business expansions.

Absent the requirement that utility operations be limited to a single integrated system, a U.S. electric and gas distribution company on the West Coast, for example, could acquire a combination gas distribution and electric company on the East Coast. Another example would be that the combination company could acquire utilities in geographically dispersed states, or buy up a group of local distribution companies (LDCs). The possibilities are endless, but nonetheless these brief examples should provide a little clearer picture on what could happen in the electric utility market.

PUHCA repeal has also opened up the development of national, multistate, or regional transmission companies. This likely will allow for greater investment opportunities and the ability to operate independently.

In the area of the purchase of nonutility businesses and expansion into nonutility enterprises, both registered and exempt holding companies are now able to purchase nonutility assets and head up expansions into unrelated lines of business. It is interesting to note that EPAct of 2005 specifically excludes banks, brokers, dealers, and underwriters of securities of their affiliates from utility holding company status under various circumstances.

Further, exempt companies that operate utility systems may now also restructure their energy trading and independent power operations. Prior to EPAct of 2005, nonutility owners of utility entities had to be creative in their utility property structuring to either avoid or minimize PUHCA's effect on those utility holdings.

In conclusion, many of the pending changes within the electric utility industry may occur with a little more ease because of the repeal of PUHCA. A great many of the changes occurring and expected to occur with the national electricity grid will be costly and will require large amounts of financial investment. Relaxation of some burdensome rules upon electric utility companies and their affiliates may make the financial viability of grid modernization more of a reality.

Transmission, Technology, and the Pursuit of Reliability

As an essential service to everyday lives, electricity—and the transmission of it to homes, business, and industry—must have certain degrees of order and means of delivery. Over the past decade as the restructuring of the electric utility industry has unfolded, reliability of the interconnected bulk electric system has become one of two central themes around which the future of the U.S. electric grid hinges.

NERC defines reliability in terms of two basic functional aspects. The first is adequacy, or the ability of the electric system to supply the aggregate electrical demand and energy requirements of customers at all times. It takes into account scheduled and reasonably expected unscheduled outages of system elements.

Secondly, according to NERC's definition, *reliability* is all about security. This is the second central theme upon which the interconnected bulk electric supply system relies upon for its future success or failure. Security is the ability of the electric system to withstand sudden disturbances such as electric short circuits or unanticipated loss of system elements.

Transmission System Status

With increased demands being placed on the transmission system as a result of electric utility industry restructuring and changing market-related needs, the nation's transmission system is being operated closer to its reliability limits than ever before.

Transmission system additions, simply put, have not kept pace with the increased demand for electricity. It is not that the electric utility industry does not see a need for transmission system improvements. Rather, the industry has held back in building new transmission structures for several reasons:

- **Financing.** Who is going to pay for the improvements?
- **Cost recovery.** How are electric utilities going to recoup their transmission system investments?
- **Siting issues.** Where should new transmission systems be built? How should that decision be made?

Transmission capability

According to NERC, in some areas of North America, increases in generating capability have surpassed the capability of the transmission system to simultaneously move all of the available electricity capable of being produced. And, as restructuring has progressed, market-based electricity transactions have increased, adding even more to grid congestion and tighter transmission operating margins.

Increased loading of the grid in and of itself has not necessarily compromised the reliability of the bulk electric power transmission grid, according to NERC, however plausible that may appear. System operators have at their disposal advanced analytical tools making it possible for them to more accurately assess transmission system conditions and thus maintain system reliability. Of course, this has not always been the case, and there is no certain, perfect way to ensure that the grid will be 100% reliable, 100% of the time.

A case in point is the August 14, 2003 blackout in the Northeast. This blackout was preceded on November 10, 1965 by a similar blackout in nearly the same area. According to the latest NERC data available, both of these events had their root cause in operator error. In its 2004 Long-Term Reliability Assessment report, NERC states that "system operators must have the appropriate skills and training and maintain the appropriate level of situational awareness to operate their systems reliably. Above all, system operators must understand and follow NERC and regional reliability standards."

In the 1965 blackout, certain parts of the grid were carrying electricity at near capacity when a shift of power flows tripped circuit breakers. At that point the shift of power and the resultant tripped breakers sent larger flows onto neighboring lines and began a chain-reaction failure. According to NERC reports, the cause was linked to an incorrectly set circuit breaker.

The 2003 blackout occurrence can be reviewed in the sidebar, "2004 Blackout Sequence of Events." According to NERC, specific chance events occurred, beginning with controllers in Ohio, where the blackout began. Reportedly, these controllers were overextended, lacked vital data, and then failed to act appropriately on outages that occurred more than one hour before the blackout.

August 14, 2004
Blackout Sequence of Events

1:58 p.m.	The Eastlake, Ohio generating plant shuts down. The plant is owned by First Energy, a company that had experienced extensive recent maintenance problems, including a major nuclear-plant incident.
3:06 p.m.	A First Energy 345-kV transmission line fails south of Cleveland, Ohio.
3:17 p.m.	Voltage dips temporarily on the Ohio portion of the grid. Controllers take no action, but power shifted by the first failure onto another power line causes it to sag into a tree at 3:32 p.m., bringing it off-line as well. While MidWest ISO and First Energy controllers try to understand the failures, they fail to inform system controllers in nearby states.
3:41 p.m. and 3:46 p.m.	Two breakers connecting First Energy's grid with American Electric Power are tripped.
4:05 p.m.	A sustained power surge on some Ohio lines signals more trouble building.
4:09:02 p.m.	Voltage sags deeply as Ohio draws 2 GW of power from Michigan.
4:10:34 p.m.	Many transmission lines trip out, first in Michigan and then in Ohio, blocking the eastward flow of power. Generators go down, creating a huge power deficit. In seconds, power surges out of the East, tripping East Coast generators to protect them. The blackout is on.

Source: NERC

Following the blackout, NERC took a number of actions to prevent and mitigate the impacts of future cascading outages:

- Instituted readiness audits of all control areas and reliability coordinators. The purpose is to identify areas for improvement as well as best practices in system operation, particularly under emergency conditions, and to help operators strive for excellence in their assigned reliability functions and responsibilities.

- Clarified reliability coordinator and control area functions, responsibilities, capabilities, and authorities through revisions to its operating policies.

- Required all reliability coordinators, control areas, and transmission operators to provide at least five days per year of training and drills in system emergencies for each staff person with responsibility for real-time operation.

- Strengthened the NERC Compliance Enforcement Program.

Additionally, as discussed in chapter 5, NERC adopted an exhaustive, lengthy new set of reliability standards. The standards, which went into effect April 1, 2005, incorporate the existing NERC operating policies, planning standards, and compliance requirements into an integrated and comprehensive set of measurable reliability standards.

Transmission grid physics

Many experts in the industry firmly believe many of the problems with the U.S. bulk electric power transmission grid stem from restructuring. Although it is difficult to pinpoint, restructuring probably has not caused as many problems as the physics of the transmission system itself. This will be examined next.

As discussed earlier, restructuring activities, which many refer to as deregulation, began in the 1990s. Prior to that time, electric utilities were fully regulated, vertical monopolies. A single company controlled electricity generation, transmission, and distribution in a given geographical area. Each utility would maintain sufficient generation capacity to meet its customers' needs. Any long-distance energy transmission services were normally confined to emergency situations. Before restructuring, the long-distance connections served only as a backstop in extreme circumstances to prevent sudden power losses for any given geographical area.

The physical behavior of the grid is such that electricity generation, transmission, and distribution that cover the United States and Canada function as a single operating unit, so to speak. This single network has three interconnects. This was discussed at length in chapter 2, but for purposes of review, a simple explanation is provided here.

The interconnects consist of the Eastern (covering the eastern two-thirds of the United States and Canada); the Western (covering most of the rest of the two countries); and the Electric Reliability Council of Texas (covering most of Texas). Within each one, power flows through AC lines, so all generators are tightly synchronized to the same 60-Hz cycle. The interconnects are joined to each other by DC links, so the coupling is much looser among the interconnects than within them. This also means that the capacity of the transmission lines between the interconnects is also far less than the capacity of the links within them.

As many readers already know, the physical complexities of power transmission rise dramatically as distance grows. To add to this, longer transmission lines have less capacity than shorter ones. Power does not travel along a set path on the electric network. As an example, when utility Y sends electricity to utility Z, utility Y increases the amount of power generated, while utility Z decreases production or has an increased demand. The power then flows from utility Y to utility Z along all the paths that can connect them. Any changes in generation and transmission at any point in the system will change loads on generators and transmission lines at every other point (and in some cases, in ways not anticipated or readily controlled).

In order to avoid system failures, the amount of power flowing over each transmission line must remain below the line's capacity. When capacity is exceeded, heat is generated in a line, which can cause the line to sag or break or can create power-supply instability such as phase and voltage fluctuations.

For an AC power grid to stay stable, the frequency and phase of all power generation units must remain synchronous within narrow limits. A generator that drops 2 Hz below 60 Hz will rapidly build up enough heat in its bearings to destroy itself. This is why when the frequency varies too much, circuit breakers trip a generator out of the system.

In the view of many industry experts, a tremendous error has been made in reformulating the rules and regulations governing the grid. The error rests in the fact that electricity is being treated as a commodity and not a service.

As discussed earlier, electricity cannot be shipped from point A to point B without causing a rippling effect. Power shifts, it has been proven, do cause the entire single machine system to act and react. In other words, unlike the commodities market, the power grid is not in a static environment.

Sell high, buy low

As discussed in chapter 4, it was the Energy Policy Act of 1992 that empowered the FERC to separate electric power generation from transmission and distribution. However, it was not until 1998 that it really took off, with California taking the lead.

The premise that has gone along with restructuring was for utilities that generated power to sell off their generating capacity to independent power producers (IPPs), such as Enron and Dynergy, for the best price they could get. Utilities would then purchase power at the least cost possible from these IPPs.

For this whole divestiture concept to work, utilities that owned transmission lines were compelled to carry power from other companies' generators in the same way as they carried their own. (This was true even if the power was being transmitted to a third party.) Adding to this, IPPs added new generating units at random locations. Often new units would be sited at locations with low labor costs, minimal local regulations, and/or attractive tax incentives. Generators added far from main consuming areas would mean that the quantity of power flows would increase exponentially. The end result was overloaded transmission lines.

The result of the new rules that have come with restructuring has been to more tightly couple the system physically and stress it closer to capacity. At the same time, it has made control more diffuse and less coordinated. From this, it appears as though blackouts should not have been so surprising.

It would become public knowledge that certain energy companies, particularly between 1998 and 2000, actually had made a game out of running transmission up to the limits of capacity. Federal investigations into Enron and other energy traders would unveil the fact that employees would file transmission schedules designed to block competitors' access to the grid and to drive up prices by creating artificial shortages. In California, this practice led to widespread blackouts, inflated retail rates, and eventual costs to ratepayers and taxpayers of more than $30 billion. There was less effect during this time period (1998–2000) in the more tightly controlled Eastern Interconnect.

Energy trading would resume its place in the marketplace in 2002 and 2003. Although power generation in 2003 increased only 3% above that in 2000, generation by IPPs doubled. Along with this has come the resultant system stress and warnings by FERC and others that trouble was again rearing its ugly head.

Major bank and investment groups such as Morgan Stanley and Citigroup began stepping in to pick up the pieces the energy trading companies, such as Enron, had left in their devastating wake of financial ruin. This has helped some, however, because trading margins have narrowed with the entrance of more players in the market. More trades are needed to pay off the huge debts incurred in buying and building generators. Many utility companies' stocks and credit ratings plummeted, and nationwide they cut 150,000 utility jobs. Undeniably, utilities do not want to relive that experience. New technology, and perhaps a better understanding now of what it will take to survive in the reordered electric utility industry, will be what determines which utilities will survive and which will not.

Transmission System Constraints

The utility industry has learned some valuable lessons. For the bulk power system to operate reliably, it must be designed and operated so that the total generation at any moment is kept equal to total electricity consumption and losses on the system. (This includes transmission and distribution.) Another hard-won lesson is that electricity should be allowed to flow through the transmission system in accordance with physical laws and not directed to flow through specific, artificial lines. And, finally, the system should be designed with reserve capacity in generation and transmission to allow for uninterrupted service when contingencies occur.

What happens to limit the system's power transfer capability? According to the EIA, there are three types of constraints:

1. Thermal/current constraints
2. Voltage constraints
3. System operating constraints

Thermal/current limitations are the most common constraints on the capability of a transmission line, cable, or transformer to carry power. The transmission line resists the flow of electrons through it, which causes heat to be produced. The actual temperatures occurring in the transmission line equipment depend on the current (the rate of flow of the electrons) and also on ambient weather conditions (such as temperature, wind speed, and wind direction). Weather affects the dissipation of the heat into the air.

Overheating can cause at least two problems. The first is that the transmission line will lose strength because the overheating reduces the expected life of the line. Second, the transmission line expands and sags in the center of each span between the supporting towers. Should the temperatures stay repeatedly high, an overhead line will permanently stretch and may cause its clearance from the ground to be less than required for safety reasons. Because overheating is a gradual process, higher current flows may be allowed for a limited amount of time. A normal thermal rating for a line is the current flow level it can support indefinitely, while emergency ratings are at levels that the line can support for specific periods only.

Voltage constraints may also be present. Voltage is a measure of the electromotive force necessary to maintain a flow of electricity on a transmission line. Fluctuations may occur due to variations in electricity demand and due to failures on transmission or distribution lines. Constraints on the maximum voltage levels are set by the design of the transmission line. If the maximum is exceeded, short circuits, radio interference, and noise may occur. As well, transformers and other equipment at the substations and/or customer facilities may be damaged or destroyed. Minimum voltage constraints also exist based on the power requirements of customers. Low voltages cause inadequate operation of a customer's equipment and may even damage motors.

It is important to note that on a transmission line, voltage tends to drop from the sending end to the receiving end. The voltage drop along the AC line is nearly proportional to reactive power flows and line reactance. Line reactance increases with the length of the line. Capacitors and inductive reactors are installed as needed on lines to (in part) control the amount of voltage drop. This is important because voltage levels and current levels determine the power that can be delivered to customers.

Finally, the bulk power system may be hampered by system operating constraints. Operating constraints stem primarily from concerns with security and reliability, which are related to maintaining the power flows in the transmission and distribution lines of a network. Power flow patterns redistribute when demands change, when generation patterns change, or when the transmission or distribution system is altered due to a circuit being switched or put out of service.

Security and System Stability

Transmission capabilities may also be greatly affected by preventive operating procedures for system security. The loss of a generating unit, transmission line, or a failure of another component of the system should not cause the bulk power system to shut down. In fact, the system is designed to keep up a continual level of service despite any losses that may occur.

NERC considers preventive operating procedures, which essentially means that the system should be run in such a way as to avoid service interruptions, to be a good utility practice. It is also the primary means of preventing disturbances in one area from causing service failures in another. NERC provides standards and operating guidelines for overall coordination of utility procedures in the United States, Canada, and parts of Mexico.

Part of NERC's preventive operating requirements include running sufficient generation capability to provide operating reserves in excess of demand and limiting transfers on the transmission system. The system then operates so that each element remains below normal thermal limits under normal conditions and under emergency limits during system disturbances. (System disturbances or contingencies—loss of generation unit, loss of transmission line, etc.—are handled using reserve capacity.)

System stability issues that arise are generally grouped into two types:

1. Maintaining synchronization among the generators of the system
2. Preventing the collapse of voltages

The interconnected operating system operates in a synchronous manner; all generators rotate in unison at a speed that produces a consistent frequency. (In the United States, this frequency is 60 cycles per second.)

Power requirements from generators are altered when a disturbance occurs in the transmission system. This may reduce the power requirements from the generator, but the mechanical power driving the turbine stays constant, and the generator accelerates.

Logically, once the disturbance is gone, the power flow once again is altered, and the turbine slows down. The result is oscillations in the speed at which the generator rotates and the frequency at which the power flows into the system. At this point, the system is unstable. Natural conditions or control systems can reduce the oscillations, but if this does not happen, the system is rendered unstable. This condition is referred to as *transient instability,* and it can lead to a complete collapse of the system. Transient instability can be avoided if power transfers between areas are limited to levels determined by system contingency studies.

Steady state instability can occur if too much power is transferred over a transmission line or part of a system to the point that the synchronizing forces are no longer effective. While steady state instability is unusual, since it is fairly easy to prevent, it can act as an impediment to transmission power transfers.

Small-signal stability can also affect power transfers adversely. Also referred to as dynamic instability, it happens when the transmission system is not designed to handle reactive power flows. Large amounts of reactive power flows on long transmission lines result in severe drops in voltage at the consumption end. This causes customers to draw increasing currents. These increased currents, in turn, will cause added reactive power flows and voltage losses in the system, thus leading to still lower voltages at the consumption end. As the process progresses, the voltages collapse further, requiring users to be disconnected to prevent serious damage. At this point, the system either partially or fully collapses.

Operating under a tight transmission system

Working with the grid that is being operated closer to its limits than ever before is most challenging for system operators. Operators must use monitoring and analysis tools more efficiently to mete out the activity on the system and the actions required of them to maintain reliability.

As a result of this tight transmission system operating environment, the electric utility industry is continually working to improve existing analytical tools. It is also introducing new ones to allow system operators to better manage the transmission system and meet user expectations. Some examples of evolving sets of new tools include the following:

- State estimators
- Wide-area monitoring systems
- Near real-time security analysis

State estimators read real-time data from thousands of points across the electrical system and use fundamental electrical engineering equations to apply this data to a model of the system. An improved set of data for the system operator to use is then generated. State estimators improve the quality of the data by obtaining a best fit between the measured quantities and their theoretical values.

However helpful, state estimators are often hampered in their ability to produce the latest data, since they rely so heavily on continuous data updates that can be adversely impacted by communication failures. As with most advanced monitoring tools, access to continuous, high-quality data is crucial. Loss of data communication from a number of critical stations, another control area, or a location that has changed states could prevent the state estimator from producing a usable solution. This is why it is important to have redundant sources of primary data that can replace data collection streams in the event that communication is interrupted.

The analysis and operation of the power grid are also heavily affected by interdependency across the interconnection. Increasing or decreasing load in one control area has an immediate impact on the power flows in other control areas. Events occurring on one portion of the bulk power system can often be observed everywhere else in the interconnection.

Information interdependency within the interconnections requires wide-area control and communications systems to enable reliable operation of the power system. Wide area monitoring systems (WAMS) are an emerging alternative that could supply broader and more comprehensive real-time information across the interconnection, according to NERC.

WAMS were first developed for Bonneville Power Administration (BPA) in the WSCC to provide for simultaneous information exchange capability. These systems are being expanded into the Eastern Interconnection under a cooperative effort of the DOE, the Consortium for Electricity Reliability Technology Solutions (CERTS), and NERC. They will provide the real-time monitoring of voltage and power and the acquisition of data necessary for the verification of planning models and programs. In addition, WAMS output can also add to the situational awareness required of system operators.

Another evolving form of analysis is near-time security analysis. State estimator results are used as a starting point to evaluate all plausible contingency scenarios. The analysis can include both single- and compound-element contingencies. System operators are alerted when the analyzer shows that a simulated post-contingency operating security limit is violated. Operators can use this information to modify the system in real time, or to develop a plan of action for possible events.

Near real-time security analysis revolves around the capabilities of computer systems and advanced software applications. In order to present a meaningful simulation, the computer model of a control area must have detailed information about the topology, loads, generation, and the line flows in its own and other control areas. This essential, yet basic, information must be shared in a timely manner between computer systems in neighboring control areas. NERC suggests that operators use information from a variety of sources to determine the state of the system, since equipment failures may occur.

Transmission line monitoring

NERC also has recommendations for transmissions owners in planning studies to determine the need for system upgrades. Static ratings should be used based on surveys that reflect a conservative estimate of the expected range of environmental and operating conditions.

In the past, transmission lines were assigned static loading limits that were determined by a fixed set of environmental or operating conditions. The type and size of conductors, and limits on other line equipment, determined the current carrying capacity of a line. No allowances were made for variations in temperature or wind speed, both of which affect the current carrying capacity of the conductor.

Dynamic line rates can now be displayed for operators in real time by employing models that factor in environmental conditions measured by sensors on the transmission line. Dynamic ratings can reduce operating costs by increasing the utilization of the existing facilities. They also can improve reliability by allowing operators to make real-time decisions and take actions based on actual field conditions. Dynamic rating schemes require several subsystems. They include sensors in the field, communication links to the office, and computer equipment to perform the advanced system analysis.

Enhancing and maintaining capability

In order to enhance or maintain the capability of the transmission system, special protection systems (SPS) have become more popular. An SPS, which is often known as a remedial action scheme, is designed to detect abnormal system conditions and initiate corrective actions to protect grid reliability and integrity.

Temporary or unplanned conditions on a transmission system often prompt the usage of an SPS. These conditions may arise from the delay of a transmission construction project, the presence of unusual system demands, or the unavailability of specific equipment during long maintenance outages. SPS can also be installed to preserve system integrity in the event of an extreme contingency, or a contingency that has such a low probability of occurrence that an investment in additional transmission would be cost prohibitive, although the consequences of the event may be severe.

Another essential maintenance tool is reactive power planning. Reactive power is the energy necessary to support the transmission of real power from generation sources to load in an AC network. Planning is thus required to ensure that sufficient reactive power reserves exist near loads to support system voltage levels.

According to NERC, due to the way reactive power is stored in and generated by magnetic and electric fields, it tends to be absorbed by heavily loaded transmission lines. It is generated by lightly loaded transmission lines. This makes reactive power difficult to transmit over long distances. As reactive power consumption in an area grows, and if no corrective action is taken, voltages decline, and the system becomes susceptible to voltage collapse. Reactive power shortages have been a contributing factor in several major blackouts over the last decade.

Reactive power sources may be either static or dynamic. Dynamic and fast-switched reactive power sources such as generators, synchronous condensers, and static VAR devices are generally considered to be sources that can respond essentially instantaneously to voltage changes. All of these reactive power sources provide immediate and automatic support for a system whenever a system contingency occurs.

Static reactive power devices, such as switched capacitor banks, are used to maintain steady-state voltage levels. Increasing the amount of reactive power available in an area can be accomplished by building new transmission lines to relieve the loading on the existing transmission grid. This will have the effect of reducing the reactive power consumption by the transmission system, which will help to support system voltage levels and real power transfer to loads. The cost of building new transmission lines (discussed later in this chapter) is high in relation to other reactive supply solutions. Thus new transmission is typically built to support an increase in real power transfer before it is considered for reactive supplies.

It is important to note that unless contractual obligations are specific, agreements between generation owners and transmission operators may not provide adequate incentives to generate reactive power. This is because real power is what is typically metered (in MW) for load payment. According to NERC, the effect of these changes is an increased loading on the existing transmission grid, as power has to be shipped into densely loaded areas from longer distances. Heavily loaded transmission lines absorb reactive power, thereby further limiting the amount of reactive power that can be transferred to load. Compounding this, of course, is the pressure being placed on transmission operators to push the grid closer to its operating limits.

NERC contends that long-term solutions to reactive power problems need to be evaluated. With increased difficulty in locating generation close to load and building new transmission, dynamic reactive reserve options could run out quickly in some of the largest load centers in the country. In addition, much attention should be given, NERC maintains, to reactive supply margin and the balance between static and dynamic reactive resources by both the planning and operations functions. This is important as systems are run more closely to their margins. NERC warns that a failure to address these issues could lead to a decrease in grid reliability.

Impact of fuel diversity

Fuel diversity is of concern relative to the impact on overall generating capacity from disruptions in the supply of different types of fuel. The general thought here is that if diversity exists in the fuels that are used for generation, a disruption in any one of the fuel chains will not have an overwhelming impact on available generation capacity.

The growing gas/electricity interdependency has brought about some heightened concerns. Natural gas is increasingly being used by the majority of new generating capacity additions. There are some issues, however, with natural gas that have come about that could negatively impact reliability. These issues include fuel diversity, fuel deliverability, and fuel availability.

In some regions, an increase in natural gas usage increases the diversity of fuel types. However in other regions, it actually reduces fuel diversity. As for fuel deliverability, in some areas deliverability to generation may be limited. Some regions of North America are only served by a few, or even one, pipeline. This alone is reason enough to factor in the operating reliability of gas infrastructure in the overall operation and planning of grid reliability.

Cyber Security

When it comes to security, it seems these days there just cannot be enough done to ensure that the U.S. transmission system is safe from terrorists, criminals, and other individuals. It is a proven fact that more than 90% of the United States' critical infrastructure companies are owned and operated by the private sector and are interconnected through networks. Among the critical infrastructure services is the energy industry.

The necessary network interconnections have been a double-edged sword. While creating efficiency, these intricate networks have also made an already vulnerable electric utility industry even more open to cyber attacks. As has been discussed in this book, the onslaught of industry restructuring has caused some hardship—and in some cases, very costly hardship. Change often has a way of creating hardship, particularly on the once-staid utility industry. However, it also has a way of preparing and strengthening the industry to new, previously unheard of levels.

NERC defined *critical cyber assets* as "those computers, including installed software and electronic data, and communication networks that support, operate, or otherwise interact with the bulk electric system operations."

So the industry is not merely experiencing security issues as they relate to preventive operating procedures for the bulk power system. It also is seeing heightened attention being given to the actual control centers, and more specifically to information technology (IT) and control networks.

The U.S. transmission grid is increasingly becoming automated. Utilities now more than ever are integrating with the grid through digital switches and other high-tech gear. These improvements are truly making the system more reliable. However, on the flip side, these improvements may be costly in terms of how that information may be used by knowledgeable hackers and other cyber terrorists.

How does this happen? There are many ways, but the most prevalent has been through worm-related attacks and denial-of-service attacks. In fact, a study was conducted by Symantec Corporation, a company that specializes in enterprise security solutions for the electric power industry. They found that between January 1, 2004 and June 30, 2004, such worms as The Slammer, MyDoom, Blaster, Welchia, and Sasser were all strongly associated with the top attacks and top targeted ports in the power and energy sector. According to Symantec, worms can have a profound impact on availability of services, productivity of employees, and the confidentiality and integrity of data, among other things. Denial-of-service attacks were also prominent in the company's findings. This type of attack significantly disrupts the availability of crucial network systems.

Some ways Symantec suggests for helping prevent cyber attacks on control networks include creating security zones by deploying integrated gateway security appliances. These include a firewall, intrusion detection/prevention, and antivirus software. All potential entry points should be secured against viruses and worms, including infected mobile computers or other computers that are connected to the utility's network.

For denial-of-service attacks, Symantec suggests that administrators evaluate the management of the security systems and consider having a managed security service provider monitor their company's cyber security continuously. Response procedures should be developed, if they are not already available, to mitigate damages from attacks. A logging policy and log aggregation policy can help utilities to speed up investigation of these attacks when they occur. In addition, Symantec recommends that security administrators ensure that patches and updates are tested and applied for all systems, and in particular, remote or mobile systems.

Challenges to preparedness for cyber threats

There are many challenges faced by utilities working to secure the power grid from cyber attacks. These include overlapping requirements from various agencies, limited access to useful information, and ineffective means of communicating identified security weaknesses from utility operations, planning, and engineering personnel to higher level management.

It is for the reasons above, as well as others, that EPRI announced its PowerSec Initiative in July 2005. According to an EPRI white paper on the new initiative:

> A number of individual utilities have performed valuable cyber security efforts, each producing useful results. However, lack of technology transfer mechanisms has limited the application of these results to the industry as a whole.

> Because security is only as strong as 'the weakest link' in the chain of interconnected information and communication systems that utilities use, broad industry support, participation, and successful broad implementation of new cyber security tools is crucial for each utility's cyber security.

With input from EPRI's board of directors and a number of utilities and industry organizations, the PowerSec Initiative is designed to assess and improve individual utility and industry preparedness for cyber attack. The initiative will be geared toward the examination of the increasing level and type of cyber threats, vulnerabilities, and potential consequences. It will use this information to evaluate current industry cyber attack readiness, identify gaps, and specify existing best practices for filling in these gaps. It is expected that the expert group's work will also include the identification of vulnerabilities that will require new solutions and specify needed research and development work to prepare and test these solutions.

According to EPRI, the initiative's first goal will be to focus on electric utility supervisory control and data acquisition (SCADA) systems and energy management systems (EMS). All utilities and organizations such as state agencies and ISOs and RTOs are being encouraged to participate.

The Role of Broadband over Power Lines

When BPL technology is discussed, it usually comes to mind as the ability to provide broadband connectivity to customers in an electric service area using electrical power lines. However, for purposes of discussion here, BPL and its function will be limited in scope to its use by electric utilities to manage their power networks more efficiently and thereby increase both security and reliability.

According to NARUC, BPL's apparent technological potential has induced several electric utilities to deploy it, if even on an experimental basis. NARUC cited the use of BPL technology by Cinergy of Cincinnati, Ohio and the municipal utility in Manassas, Virginia as the most notable examples.

Cinergy, which may have the most extensive rollout of BPL-enabled broadband service to electric customers, according to NARUC, has its sights set on BPL's potential for improving electric service. The city of Manassas, Virginia has its primary vision on utilizing BPL to enhance its system-monitoring capabilities. Through making available high-speed Internet access to its customers, Manassas can pinpoint where faults and failures are occurring in much finer geographic detail. The city also has reportedly converted from manual to automated meter readings to cut costs and improve accuracy.

Other examples come from Consolidated Edison Company and Hawaiian Electric Company. These electric IOUs have rolled out BPL pilot projects to improve their operational capabilities.

The Broadband Over Power Lines Task Force of NARUC was formed in 2003 to explore BPL's potential for the industry. A report issued by the task force noted that BPL has tremendous potential for enhancing the operability of the electric grid. The report goes on to say that the ultimate goal is the development of the "intelligent" or "smart" grid.

This smart grid refers to an electricity transmission and distribution system that incorporates elements of traditional and cutting-edge power engineering, sophisticated sensing and monitoring technology, IT, and communications. These will provide better grid performance and support a wide array of additional services to customers. Table 7–1 details key attributes of the 21st-century grid.

Table 7–1. Key attributes of the 21st-century smart grid

Grid Attribute	Potential Enhancements
Self-healing	Sophisticated grid monitors and controls will anticipate and instantly respond to system problems in order to avoid or mitigate power outages and power quality problems.
Secure	Deployment of new technology will allow better identification and response to both physical as well as cyber threats.
Standardized	The grid will support widespread use of distributed generation. Standardized power and communication interfaces will allow customers to interconnect fuel cells, renewable generation, and other distributed generation on a simply "plug and play" basis.
Consumer Friendly	Consumers will be able to better control the appliances and equipment in their homes and businesses. The grid will interconnect with energy management systems in smart buildings to enable customers to manage their energy use and reduce their energy costs.
Efficient	Grid upgrades that increase the throughput of the transmission grid and optimize power flows will reduce waste and maximize use of the lowest-cost generation resources. Better harmonization of the distribution and local load servicing functions with interregional energy flows and transmission traffic will also improve utilization of existing system assets.

Source: NARUC

While the smart grid remains a futuristic concept, NARUC's report stresses how the smart grid could enhance security and reliability. It will be interesting to watch this up-and-coming technology be used to enhance not only power delivery but also to provide significant benefits to consumers.

Transmission System Upgrades

The question is not what has happened that will cause an aggressive course of action to upgrade the U.S. bulk power transmission system, but rather what will happen if no action is taken at all.

Investment in transmission system upgrades has been on the burner (if even the back burner) since the August 2003 blackout. There remains a plethora of ideas on how, when, and where to make the necessary upgrades, but no significant resultant action.

One of the biggest hindrances has been the cost, and the question of who will pay for the upgrades. According to the Cambridge Energy Research Associates (CERA) study, *Grounded in Reality: Bottlenecks and Investment Needs of the North American Transmission System*, May 2004, this financial conundrum is resulting in persistent transmission constraints in the eastern three-fourths of the United States. This equates to around $300 million of potential annual energy cost savings that CERA claims could be realized from much-needed congestion relief projects. These projects continue to go unfunded.

The most cost-effective solutions, according to CERA's report, are to move power from lower-cost generation sources in the Midwest and Southeast to load centers on the East Coast and in Florida and the southern Great Plains. This would be done by building new transmission lines and upgrading equipment along transmission routes. The total price tag for these enhancements was cited by CERA at about $2.2 billion by 2010. This amount would be in addition to the investment currently under way of around $2 billion to $4 billion annually in the United States. This current investment is just to maintain the existing transmission system, meet local reliability needs, and provide new generator interconnections.

Build or build on transmission lines

There exist several avenues in which constraints on the existing transmission system can be remedied. Due to the tremendous costs of building new transmission lines, a lot of space will not be devoted here to discuss this remedy. However, it does merit at least a few words.

As shown in table 7–2, the typical cost for new transmission lines is quite expensive. It should be noted that the actual costs for a specific project could be somewhat higher or lower than that shown in table 7–2. For example, some costs are not included in the amounts shown in the table, because they vary so widely from location to location. These extra costs include right-of-way (ROW) costs or the cost of land and the legal right to use and service the land on which the transmission line would be located. Nonetheless, the point is that in most cases, new transmission lines are often built as a last resort, when no other remedy will address a given problem.

Table 7–2. Typical costs and capacity of new transmission lines

Voltage	Type of Supporting Tower and Number of Circuits	Size of Power Line	Normal Rating MW	Cost per Circuit per Mile[a]
		Above Ground		
60 kV	wood pole, single	4/0 AWG	32	$120,000
60 kV	wood pole, single	397.5 kcmil	56	$125,000
60 kV	wood pole, single	715.5 kcmil	79	$130,000
115 kV	wood pole, single	4/0 AWG	64	$130,000
115 kV	wood pole, single	397.5 kcmil	108	$135,000
115 kV	wood pole, single	715.5 kcmil	151	$140,000
115 kV	steel pole, single	715.5 kcmil	151	$250,000
115 kV	steel pole, single	715.5 kcmil, bundled	302	$400,000
115 kV	steel pole, double	715.5 kcmil	151	$160,000
115 kV	steel pole, double	715.5 kcmil, bundled	302	$250,000
230 kV	steel pole, single	1,113 kcmil	398	$360,000
230 kV	steel pole, single	1,113 kcmil, bundled	796	$530,000
230 kV	steel pole, single	2,300 kcmil, bundled	1,060	$840,000
230 kV	steel pole, double	1,113 kcmil	398	$230,000
230 kV	steel pole, double	1,113 kcmil, bundled	796	$350,000
230 kV	steel pole, double	2,300 kcmil, bundled	1,060	$550,000
		Underground		
115 kV	underground cable	200 MVA	180	$3,300,000
230 kV	underground cable	400 MVA	360	$3,700,000

[a]These costs do not include right-of-way costs.
AWG = American wire gauge.
kcmil = One kcmil is 1,000 circular mils, a measure of wire cross-area.
kV = Kilovolts.
MVA = Megavolt amperes.
MW = Megawatts.
Source: Energy Information Administration

According to the EIA, there are many options available for reducing the limitations on power transfers due to the thermal rating constraints on overhead transmission lines. The thermal limit of a transmission line is based on the component that would be the first to overheat. Thus a substantial increase in the overall thermal rating of the line can sometimes be found in replacing an inexpensive element. The replacement of a disconnect switch or circuit breaker is much less costly than major work to replace a line or to build a new one.

Another remedy for thermal constraints is to increase allowable temperatures and plan for a decrease in the life of the lines. It may also be possible to increase the transfer capability of the line by monitoring the line sag to allow higher temperatures/currents.

A third remedy, which ranks next to building new lines in terms of cost, is to replace the lines with large ones through restringing or adding one or more lines, forming bundled lines. This approach will require consideration of the tower structures that support the power lines. The towers are designed to hold the weight of the existing lines and the weight of any possible ice formations. So, substantial reinforcement of the towers may be required as well as an enhancement of substation equipment to support the restringing.

To address the issue of voltage constraints for individual lines, EIA states that increasing the operating voltage within a voltage class is a technique that has been used for decades. This is a good and viable option provided that the system does not reach the upper voltage limit during light loads under normal operation. Voltages of the generators will need to be increased and some adjustments to the settings of the transformers will also need to occur to produce the new operating voltage. In addition, coordination with neighboring systems would be required to prevent additional reactive power flows because of the increased voltage into the neighboring system.

According to EIA, other remedies for voltage problems that limit transfer capabilities involve controlling reactive power flows. Capacitors and reactors are the two types of reactive power sources, and they generate and absorb reactive power flows, respectively. The installation of capacitors or reactors at strategic locations of the transmission or distribution system is a common remedy used to control reactive power flows and therefore increase power transfers.

Other methods of mitigating power transfer constraints due to individual components include converting single-circuit towers to multiple-circuit towers and converting AC lines to HVDC. According to EIA, most transmission circuits for 230 kV and below are built on two-circuit tower lines. Circuits for higher voltages are generally built on single-circuit towers. Substantial increases in either ROW width or in tower height are required for conversion of a single-circuit line to a double-circuit line. Estimates of the costs of conversion are shown in table 7–3.

Table 7–3. Estimates for converting single-circuit tower lines to double-circuit lines

Conversion to	Cost per Mile (New Tower Assembly Not Required)	Cost per Mile (New Tower Assembly Required)
60 kV (397.5 kcmil, unbundled)	$40,000	NA
115 kV (715.5 kcmil, unbundled)	$80,000	$320,000
115 kV (715.5 kcmil, bundled)	$130,000	$500,000
230 kV (1,113 kcmil, unbundled)	$120,000	$460,000
230 kV (1,113 kcmil, bundled)	$200,000	$700,000
230 kV (2,300 kcmil, bundled)	$260,000	$1,100,000

kcmil = One kcmil is 1,000 circular mils, a measure of wire cross-area.
kV = Kilovolts.
NA = Not applicable.

Source: Energy Information Administration

Converting an AC line to HVDC or replacing an AC line is a consideration when large amounts of power are transmitted over long distances. HVDC lines are connected to AC systems through converter systems at each end. The power is converted from AC to DC at the sending end and back to AC at the receiving end. HVDC circuits have some advantages over AC circuits for transferring large amounts of power. HVDC circuits can be controlled to carry a specific amount of power without regard to the operation of the AC circuits to which they are connected. If HVDC lines are operating in parallel with AC lines, the outage of a parallel AC line does not overload the DC line, but the outage of the HVDC line does increase the loading on the parallel AC lines. HVDC circuits have resistance but do not have reactance associated with AC, so they have less voltage drop than AC circuits.

The major disadvantage to HVDC circuits is that they require converter stations at each end of the circuit that are expensive. This fact makes HVDC uneconomical except when power is being transmitted over long distances.

Meeting the challenge of system operating constraints may entail a rather simplistic approach. Sometimes the power flows through a transmission system can be improved by changing the connections of lines at various substations to increase power flow through some lines and reduce it in others. Some reconfigurations, such as closing some circuit breakers and opening others, require little or no monetary investment.

There has also been a shift in operating philosophies that bears mention here. The technologies of today are leaning more toward corrective, rather than preventive, methods. While corrective operation tends to be less reliable than preventive operation, corrective operation allows greater power transfers during normal operations.

According to EIA, technologies developed as part of FACTS can be used to help mitigate current preventive system operating constraints. A FACTS device can be used to reduce the flow on the overloaded line and increase the utilization of the alternative paths' excess capacity. This allows for increased transfer capability in existing transmission and distribution systems under normal conditions.

So it stands to reason that as restructuring continues, future operators of the transmission system—ISOs, RTGs, power pools, and utilities—may want to increase the utilization rates of existing transmission lines first before investing in new building activity. At least for the time being, this appears to be how the industry is leaning. As discussed earlier, a lot of the concern is over who would pay for the new lines, how that cost would be recouped, and over what time period. With all the uncertainties, the industry may continue to see creative ways to build onto, rather than build, new transmission lines.

Environmental Standards and Issues

8

The issues of clean air and air pollution are inextricably linked to the electric utility industry. While there are numerous environmental rules and regulations imposed upon the industry, the focus here will be contained to a discussion of air quality standards. The purpose is not to point a finger at electric utilities for their almost inescapable part in these issues. Rather it is to discuss the history behind clean air rules and what the future may hold for the significant improvement of air quality.

Clean air and air pollution has been the subject of much public debate for centuries. In 1306, King Edward I of England issued a proclamation banning the use of sea coal in London due to the smoke it caused. Over the course of the next few centuries, several efforts were made in the Great Britain to reduce the amount of smoke in the air.

Clean Air Rules

The first attempt at controlling air pollution in the United States occurred during the industrial revolution. Chicago and Cincinnati enacted clean air legislation in 1881. At one point, the Bureau of Mines, under the Department of the Interior, created an Office of Air Pollution to control smoke emissions. That office no longer exists. It was eliminated shortly after its creation due to inactivity.

During the late 1940s, serious smog incidents in Los Angeles and Donora, Pennsylvania raised public awareness and concern about air pollution. In 1955, the government deemed the issue worthy of being dealt with on a national level. So, in 1955 the Air Pollution Control Act, the first in a series of clean air and air quality control acts, was enacted. These acts continue to be in force today, and they have been revised and amended over the years.

This chapter will examine the latest occurrences specifically related to the Clean Air Rules of 2004. These exhaustive mandates are a suite of interrelated rules that address ozone and fine particle pollution, and power plant emissions of sulfur dioxide, nitrogen oxides, and mercury. The purpose of the rules is to provide national tools to achieve significant improvements in air quality and the associated benefits. In fact, the rules have been touted as the beginning of the one of the most productive periods of air quality improvement in America's history.

Clean Air Interstate Rule

On March 10, 2005, the EPA issued the Clean Air Interstate Rule (CAIR), a rule it contends will achieve the largest reduction in air pollution in more than a decade (see fig. 8–1).

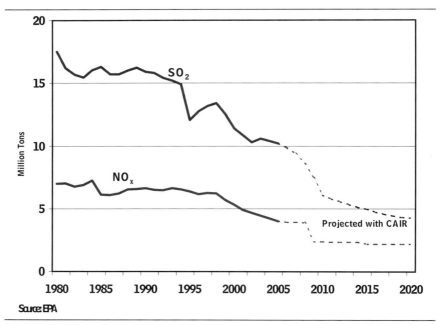

Fig. 8–1. CAIR accelerates 35 years of clean air progress: nationwide SO_2 and NO_x emissions from the power sector

CAIR specifically provides states with a solution to the problem of power plant pollution that drifts from one state to another. It will permanently cap emissions of sulfur dioxide (SO_2) and nitrogen oxides (NO_x) in the eastern United States. According to EPA, when fully implemented, CAIR will reduce SO_2 emissions in some 28 eastern states and the District of Columbia by more than 70%. NO_x emissions are expected to be reduced by 60% from 2003 levels.

Some may wonder why SO_2 and NO_x emissions are regulated. Public health studies have found that these two emissions contribute to the formation of fine particles, and NO_x contributes to the formation of ground-level ozone. Fine particles and ozone are associated with thousands of premature deaths and illnesses each year. In addition, these pollutants reduce visibility and damage sensitive ecosystems.

CAIR establishes a cap-and-trade system for SO_2 and NO_x based upon EPA's proven Acid Rain Program. The Acid Rain Program has reduced SO_2 emissions faster and cheaper than anticipated, according to EPA, and has resulted in widespread environmental improvements.

Under this cap-and-trade approach in CAIR, EPA will allocate emission allowances for SO_2 and NO_x to each state. States will then distribute those allowances to affected sources, which can trade them. According to EPA, as a result, sources will be able to choose from any compliance alternatives. Several are listed here:

- Installing pollution control equipment
- Switching fuels
- Buying excess allowances from other sources that have reduced their emissions

Because each source must hold sufficient allowances to cover its emissions each year, the limited number of allowances available ensures required reductions are achieved. Furthermore, the flexibility of allowance trading creates financial incentives for electricity generators to look for new and low-cost ways to reduce emissions and improve the effectiveness of pollution control equipment.

According to EPA, the mandatory emission caps, stringent emissions monitoring, and reporting requirements with significant automatic penalties for noncompliance ensure that human health and environmental goals are achieved and sustained.

The role of states in gaining compliance with CAIR is noteworthy. First, states must achieve the required emission reductions. This may be done by meeting the state's emission budget through requiring power plants to participate in the EPA-administered cap-and-trade system. It could also be accomplished by meeting an individual state emissions budget through measures of the state's choosing.

Second, CAIR provides a federal framework that requires states to reduce emissions of SO_2 and NO_x. EPA anticipates that states will achieve this primarily through reducing emissions from the power generation sector. These reductions will likely be substantial and cost-effecting, EPA notes, so in many areas the reductions are large enough to meet the air quality standards.

Mercury Rule

A closely related action to the CAIR is the Clean Air Mercury Rule (CAMR). It is the first ever federally mandated requirement that coal-fired electric utilities reduce their emissions of mercury. It makes the United States the first country in the world to regulate mercury emissions from utilities. Together, CAMR and the CAIR create a multipollutant strategy to reduce emissions throughout the United States.

Specifically, CAMR builds on CAIR to significantly reduce emissions from coal-fired power plants. When fully implemented, EPA estimates that these rules will reduce utility emissions of mercury from 48 tons per year to 15 tons. This represents a reduction of nearly 70% (see fig. 8–2).

CAMR establishes certain standards of performance, limiting mercury emissions from new and existing coal-fired power plants. The rule also creates a market-based cap-and-trade program that will reduce nationwide utility emissions of mercury in two phases:

- **Phase I.** A 38-ton cap will be imposed with emissions being reduced through taking advantage of co-benefit reductions (i.e., mercury reductions achieved by reducing SO_2 and NO_x emissions under CAIR).

- **Phase II.** Coal-fired power plants will be subject to a second cap, which will reduce emissions to 15 tons upon full implementation. This second phase is due to be put into effect in 2018.

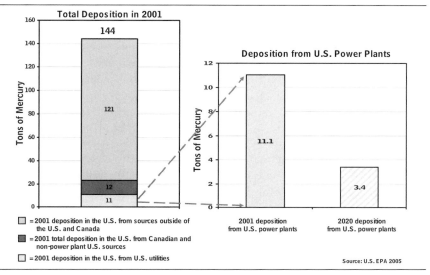

Fig. 8–2. Mercury deposition in the United States

Another requirement is that all new coal-fired plants under construction starting on or after January 30, 2004 have to meet stringent new source performance standards in addition to being subjected to the caps.

CAMR's cap-and-trade system for mercury is also based upon EPA's Acid Rain Program. Under CAMR, EPA has assigned each state and two tribes an emissions budget for mercury, with each state being required to submit a plan detailing how it will meet its budget for reducing mercury from coal-fired plants. As is similar under CAIR, states may join the trading program by adopting the model trading rule in state regulations, or they may adopt regulations that mirror the necessary components of the model trading rule.

According to EPA, the mandatory declining emissions caps, coupled with significant penalties for noncompliance, will help to ensure that the rule's mercury reduction requirements are achieved and sustained. Additionally, stringent emission monitoring and reporting requirements will help ensure that the monitored data is accurate, that reporting is consistent among sources, and that the emission reductions occur.

Facts about mercury

When coal is burned, mercury is emitted. While coal-fired plants are the largest remaining source of human-generated mercury emissions in the United States, they contribute very little to the global mercury problem. It is interesting to note that coal-fired plants emit mercury in three different forms—oxidized mercury, elemental mercury, and particulate mercury.

Estimates of annual total global mercury emissions from all sources (both natural and human-generated) range from roughly 4,400 to 7,500 tons per year. Human-caused mercury emissions are estimated to account for roughly 3% of the global total; U.S. coal-fired power plants are responsible for only about 1%.

EPA has conducted extensive tests on mercury emissions from coal-fired plants and subsequent regional patterns of deposition to U.S. waters. Those analyses conclude that regional transport of mercury emission from coal-fired power plants in the United States is responsible for very little of the mercury in U.S. waters. That contribution, however minimal, will be significantly reduced after CAIR and CAMR are implemented.

Ozone Rules

The 8-Hour Ground-level Ozone Designations are also part of the overall Clean Air Rules of 2004. A *designation* is the term EPA uses to describe the air quality in a given area for any of six common pollutants, known as criteria pollutants. These pollutants, which include ground-level ozone, are unhealthy to breathe. Vehicle exhaust and industrial emissions, such as those from power plants, are major sources of NO_x and volatile organic compounds that help to form ozone.

EPA designates an area as nonattainment if it has violated (or has contributed to violation of) the national eight-hour ozone standard over a three-year period. An area may also be designated as attainment/unclassifiable if it has monitored air quality data showing that area has not violated the ozone standard over a three-year period. It also may receive this designation if there is not enough information to determine the air quality in the area.

Fine Particle Rules

Particle pollution is a mixture of microscopic solids and liquid droplets suspended in air. This pollution is also known as particulate matter, and it is comprised of a number of components. These include acids (i.e., nitrates and sulfates), organic chemicals, metals, soil or dust particles, and allergens (i.e., fragments of pollen or mold spores).

Fine particle pollution, or PM2.5, is particulate matter that is less than or equal to 1/30th the diameter of a human hair. It can be emitted directly or formed secondarily in the atmosphere. Sulfates are a type of secondary particle formed from SO_2 emissions from power plants and industrial facilities. Nitrates, another type of fine particle, are formed from emissions of NO_x from power plants, automobiles, and other combustion sources. The exact chemical composition of particles depends upon location, time of year, and weather.

Health studies have shown a significant association between exposure to fine particles and premature death from heart or lung disease. Fine particles can aggravate heart and lung diseases and have been linked to effects such as cardiovascular symptoms, cardiac arrhythmias, heart attacks, respirator symptoms, asthma attacks, and bronchitis.

EPA issued the fine particles standards in 1997 after evaluating hundreds of health studies and conducting an extensive peer review process. The annual standard is a level of 15 micrograms per cubic meter, based on the three-year average of annual mean PM2.5 concentrations. The 24-hour standard is a level of 65 micrograms per cubic meter, determined by the three-year average of the annual 98th percentile concentrations.

Ultra-Clean Power Plants

Could it be possible for power plants to be virtually pollution free? Thanks to advancements in technology, there are several coal-fired electric utilities being proposed across the country that would utilize state-of-the-art pollution controls that hold the promise of sequestering their carbon dioxide emissions.

A March 30, 2005 report by the DOE's Technology Laboratory states that 114 coal plants capable of generating more than 70,387 MW of electricity are in various stages of development. What is even more interesting is the fact that the companies doing the building of these new plants are considering the incorporation of IGCC pollution control technologies.

The IGCC process, rather than burning coal directly, breaks the mineral down into its chemical constituents and collects each by-product for filtering or treatment before combustion. The by-products include carbon dioxide, sulfur dioxide, nitrogen oxide, trace metals, and particulates.

According to DOE, IGCC technology could be a stepping stone toward industrial-scale carbon sequestration, which is the process of capturing CO_2 and storing it in underground geologic formations.

If all 114 plants, scheduled to be built in such states as Illinois, Kentucky, Ohio, Pennsylvania, and others, are brought on line, the industry is set to spend a total of $92 billion on the new plants and the associated IGCC technology.

The Electric Utility Industry as a Business Enterprise

9

Now more than ever, the atmosphere within the electric utility industry is a stronger-than-ever movement toward looking at the industry as a business—and not just any business, but a service enterprise. Industry restructuring has brought to bear new responsibilities and demands. It has also resulted in a heightened awareness that in order to survive and thrive in the industry, electric utilities must be lean, they must be aggressive, and they must be able to deliver.

Electric utilities in the United States are classed according to their operation. The classifications include investor-owned, federally owned, publicly owned, and cooperatively owned utilities. For all intents and purposes, it does not matter what classification a utility falls into—the fact remains that there is a new playing field now and all must be operating in a more competitive, more attentive approach to the delivery of their service—electricity.

This chapter is a fitting end to the various topics of discussion contained herein. For it is in the business enterprises themselves that the success or failure of the industry rests. There is a nationwide trend toward bigger regional players, who are hopefully more efficient and effective in providing the essential service of electricity. This will most likely become the norm in the months and years ahead as companies seek to specialize and combine forces with other players to complement and grow their own core businesses.

Mergers and Acquisitions

While the electric utility industry experienced a slowdown in M&A activity during 2002 and 2003, the old adage, "out of sight, out of mind" definitely did not hold true.

The year 2004 was a banner year for M&A activity. All totaled, $50 billion worth of unions among electric and natural gas utilities were announced in 2004, according to Thomson Financial. This represents the third-highest annual amount in the last 15 years, exceeded only by the M&A-rich years of 1998 and 1999.

One of the largest deals was the $13 billion megadeal between Exelon and Public Service Enterprise Group. This pairing, announced in late December 2004, will create the nation's largest utility. The new entity, Exelon Electric & Gas, will provide power to 7 million customers and natural gas to 2 million in Illinois, Pennsylvania, and New Jersey. All told, Exelon Electric & Gas will serve a territory comprised of 18 million people.

Some may wonder why M&A activity has experienced such a comeback. The slowdown in M&A activity in 2002 and 2003 was largely due to the fact that utilities were carrying excessive debt and were concerned with having their credit ratings lowered. The business fiascos of Enron, Dynegy, and El Paso, coupled with rising gas prices, also created an instability that most utilities wanted to avoid.

With an improving economy and a relatively stable stock market, 2005 and beyond promise to be big M&A years, according to industry experts. These same experts postulate that the industry may see such companies as Florida Power & Light, Southern Company, and Duke Energy step up to the M&A plate. Already, Duke Energy has announced its intentions of merging with Cinergy. Ultimately, the expectation is that vertically integrated companies that are involved in transmission, generation, and distribution will diminish over time. This will leave more specialized utilities (in just one area) combining with other companies specializing in different functions to help their businesses become more effective.

Energy Service Companies Rise to the Occasion

Energy service companies (ESCOs), while not new to the electric utility industry, have lately become a necessary staple in the industry. Their continued presence is proving that ESCOs have a tremendous role to fill as utilities seek to provide reliable and environmentally sound energy. And, the intensifying of M&A activity is fueling ESCOs worth even more.

ESCOs were launched as a result of the 1977 Energy Bill that offered tax credits for energy efficiency. However, the role of ESCOs has evolved dramatically over the years. Their multibenefit portfolio includes helping to bring about long-term reductions in energy consumption, benefiting all ratepayers. These efforts have also relieved congestion of transmission and distribution lines and precluded the need for power generation new construction.

In 2005, ESCOs have actually seen resurgence in the use of their many services. There was a slowdown prior to 2005 primarily because energy-saving performance (ESP) contracts, a staple of the industry, had undergone federal government cutbacks. However, the Energy Policy Act of 2005 has restored ESP contracts, and thus, according to industry experts, the rekindled interest in ESCOs.

The future will likely see electric utilities hiring ESCOs for demand response or load control. ESCOs, of which the largest are owned by Honeywell, Siemens Building Technologies, and Johnson Controls, are keenly tuned in to helping utilities to combine energy efficiency and demand response in the same program.

Another new area of revenue for ESCOs is coming from renewable resources. The nonprofit Conservation Services Group, based in Westborough, Mass., has designed and built photovoltaic power plants for AEP in Texas, the Department of Defense, and Massachusetts Technology Collaborative. Chevron Energy Solution's large solar project at the Santa Rita Correction Facility is another example of this trend.

Another new ESCO market is under consideration. The Kyoto Protocol Agreement has set a ceiling on CO_2 emissions from country to country. Thus utilities that want to build a new power plant must reduce emissions through energy efficiency or through buying a credit. This has got many ESCOs looking at beginning an emissions trading market, where emissions credits can be traded.

Conclusion

While it would be impossible to sum up the state of the electric utility industry in a nutshell, there does come to mind this example and series of thoughts. Stricken with the ethical bug, as more and more utility industry executives ought to be, Sempra Energy's Neal E. Schmale serves as an excellent example of what is important in running a business enterprise.

Sempra, a Fortune 500 energy services company, serves more than 29 million consumers worldwide and has more than 13,000 employees. One would think that its leader, Mr. Schmale, would be overwhelmed. This is not the case. Instead, this executive vice president, CFO, and soon to be COO, ranks running an ethical workplace as an "imperative."

Mr. Schmale actually stated in an interview with a national magazine that the drivers of his company's success stem first from being ethical, fair, and honest. In an age where it is not uncommon to see so many ethically lax and financially wrecked companies, this is refreshing.

The second driver he listed was Sempra's management team being able to challenge conventional wisdom. Mr. Schmale said he and his team trusted their instincts about changes in the natural gas market and have benefited from that by assuming a prolific role in the emerging liquefied natural gas business.

The third success driver, and everyone can learn from this one, was flexibility. Mr. Schmale indicated that whatever assumptions one might make about the future, the only certain thing is that at least some of those assumptions will be wrong. Flexibility has allowed Sempra to be able to change course quickly to cope with regulatory, political, and market developments.

Above all else in the electric utility industry, a passion for ethical behavior, a desire for thinking outside the box now and then, and the capability of handling things as they come with poise and flexibility seem to be pretty good models for securing the future of the industry.

One could learn a lot from Mr. Schmale.

Appendix

Glossary

A

account classification. The way in which suppliers of electricity, natural gas, or fuel oil classify and bill their customers. Commonly used account classifications are *residential, commercial, industrial,* and *other*. Suppliers' definitions of these terms vary from supplier to supplier. In addition, the same customer may be classified differently by each of its energy suppliers.

accounting system. A method of recording accounting data for a utility or company or a method of supplying accounting information for controlling, evaluating, planning, and decision making.

acid mine drainage. Water pollution that results when sulfur-bearing minerals associated with coal are exposed to air and water and form sulfuric acid and ferrous sulfate. The ferrous sulfate can further react to form ferric hydroxide, or yellow boy, a yellow-orange iron precipitate found in streams and rivers polluted by acid mine drainage.

acid rain. Also called acid precipitation or acid deposition, acid rain is precipitation containing harmful amounts of nitric and sulfuric acids formed primarily by sulfur dioxide and nitrogen oxides released into the atmosphere when fossil fuels are burned. It can be wet precipitation (rain, snow, or fog) or dry precipitation (absorbed gaseous and particulate matter, aerosol particles, or dust). Acid rain has a pH below 5.6. Normal rain has a pH of about 5.6, which is slightly acidic. The term pH is a measure of acidity or alkalinity and ranges from 0 to 14. A pH measurement of 7 is regarded as neutral. Measurements below 7 indicate increased acidity, while those above indicate increased alkalinity.

active solar. As an energy source, energy from the Sun is collected and stored using mechanical pumps or fans to circulate heat-laden fluids or air between solar collectors and a building.

actual peak reduction. The actual reduction in annual peak load (measured in kilowatts) achieved by customers who participate in a utility demand-side management (DSM) program. It reflects the changes in the demand for electricity resulting from a utility DSM program that is in effect at the same time the utility experiences its annual peak load, as opposed to the installed peak load reduction capability (i.e., potential peak reduction). It should account for the regular cycling of energy-efficient units during the period of annual peak load.

adjustable speed drives. Drives that save energy by ensuring the motor's speed is properly matched to the load placed on the motor. Terms used to describe this category include polyphase motors, motor oversizing, and motor rewinding.

adjusted electricity. A measurement of electricity that includes the approximate amount of energy used to generate electricity. To approximate the adjusted amount of electricity, the site-value of the electricity is multiplied by a factor of 3. This conversion factor of 3 is a rough approximation of the BTU value of raw fuels used to generate electricity in a steam-generation power plant.

adjustment bid. A bid auction conducted by the ISO or power exchange to redirect supply or demand of electricity when congestion is anticipated.

administrative and general expenses. Expenses of an electric utility relating to the overall directions of its corporate offices and administrative affairs, as contrasted with expenses incurred for specialized functions. Examples include office salaries, office supplies, advertising, and other general expenses.

adverse water conditions. Reduced stream flow, lack of rain in the drainage basin, or low water supply behind a pondage or reservoir dam resulting in a reduced gross head that limits the production of hydroelectric power or forces restrictions to be placed on multipurpose reservoirs or other water uses.

affiliate. An entity which is directly or indirectly owned, operated, or controlled by another entity.

aggregator. Any marketer, broker, public agency, city, county, or special district that combines the loads of multiple end-use customers in negotiating the purchase of electricity, the transmission of electricity, and other related services for these customers.

air pollution abatement equipment. Equipment used to reduce or eliminate airborne pollutants, including particulate matter (dust, smoke, fly ash, dirt, etc.), sulfur oxides, nitrogen oxides (NO_x), carbon monoxide, hydrocarbons, odors, and other pollutants. Examples of air pollution abatement structures and equipment include flue-gas particulate collectors, flue-gas desulfurization units, and nitrogen oxide control devices.

air-conditioning intensity. The ratio of air-conditioning consumption or expenditures to square footage of cooled floor space and cooling degree-days (base 65°F). This intensity provides a way of comparing different types of housing units and households by controlling for differences in housing unit size and weather conditions. The square footage of cooled floor space is equal to the product of the total square footage times the ratio of the number of rooms that could be cooled to the total number of rooms. If the entire housing unit is cooled, the cooled floor space is the same as the total floor space. The ratio is calculated on a weighted, aggregate basis according to this formula: Air-Conditioning Intensity = BTU for Air Conditioning/(Cooled Square Feet x Cooling Degree-Days).

allowance for funds used during construction. A noncash item representing the estimated composite interest costs of debt and a return on equity funds used to finance construction. The allowance is capitalized in the property accounts and is included in income.

alternating current (AC). An electric current that reverses direction at regular intervals, having a magnitude that varies continuously in a sinusoidal manner.

alternative energy. Alternative fuels that are transportation fuels other than gasoline and diesel, even when the type of energy, such as natural gas, is traditional; also the use of traditional energy sources, such as natural gas, in untraditional ways, such as for distributed energy at the point of use through microturbines or fuel cells; and future energy sources, such as hydrogen and fusion.

alternative fuel. Alternative fuels, for transportation applications, including the following: methanol, denatured ethanol, and other alcohols, fuel mixtures containing 85% or more by volume of methanol, denatured ethanol, and other alcohols with gasoline or other fuels—natural gas, liquefied petroleum gas (propane), hydrogen, coal-derived liquid fuels, fuels (other than alcohol) derived from biological materials (biofuels such as soy diesel fuel), electricity (including electricity from solar energy); The United States Code, Title 42, defines it as "any other fuel the Secretary determines, by rule, is substantially not petroleum and would yield substantial energy security benefits and substantial environmental benefits." The term *alternative fuel* does not include alcohol or other blended portions of primarily petroleum-based fuels used as oxygenates or extenders, i.e., MTBE, ETBE, other ethers, and the 10% ethanol portion of gasohol.

alternative-fuel vehicle (AFV). A vehicle designed to operate on an alternative fuel (e.g., compressed natural gas, methane blend, or electricity). The vehicle could be either a dedicated vehicle designed to operate exclusively on alternative fuel or a nondedicated vehicle designed to operate on alternative fuel and/or a traditional fuel.

alternative-rate DSM program assistance. A DSM program assistance that offers special rate structures or discounts on the consumer's monthly electric bill in exchange for participation in DSM programs aimed at cutting peak demands or changing load shape. These rates are intended to reduce consumer bills and shift hours of operation of equipment from on-peak to off-peak periods through the application of time-differentiated rates. For example, utilities often pay consumers several dollars a month (refund on their monthly electric bill) for participation in a load control program. Large commercial and industrial customers sometimes obtain interruptible rates, which provide a discount in return for the consumer's agreement to cut electric loads upon request from the utility (usually during critical periods, such as summer afternoons when the system demand approaches the utility's generating capability).

amortization. The depreciation, depletion, or charge-off to expense of intangible and tangible assets over a period of time. In the extractive industries, the term is most frequently applied to mean either (1) the periodic charge-off to expense of the costs associated with nonproducing mineral properties incurred prior to the time when they are developed and entered into production or (2) the systematic charge-off to expense of those costs of productive mineral properties (including tangible and intangible costs of prospecting, acquisition, exploration, and development) that had been initially capitalized (or deferred) prior to the time the properties entered into production, and thereafter are charged off as minerals are produced.

ampere (A). The unit of measurement of electrical current produced in a circuit by 1 volt acting through a resistance of 1 ohm.

ancillary services. Necessary to support the transmission of energy from resources to loads while maintaining reliable operation of the transmission provider's transmission system. Examples include voltage control, reactive power dispatch, spinning reserve, quick-start reserve, and load following.

annual operating factor. The annual fuel consumption divided by the product of design firing rate and hours of operation per year.

anthracite. The highest rank of coal; used primarily for residential and commercial space heating. It is a hard, brittle, and black lustrous coal, often referred to as hard coal, containing a high percentage of fixed carbon and a low percentage of volatile matter. The moisture content of fresh-mined anthracite generally is less than 15%. The heat content of anthracite ranges from 22 to 28 million BTU per ton on a moist, mineral-matter-free basis. The heat content of anthracite coal consumed in the United States averages 25 million BTU per ton, on the as-received basis (i.e., containing both inherent moisture and mineral matter). Since the 1980s, anthracite refuse or mine waste has been used for steam electric power generation. This fuel typically has a heat content of 15 million BTU per ton or less.

anthropogenic. Made or generated by a human or caused by human activity. The term is used in the context of global climate change to refer to gaseous emissions that are the result of human activities, as well as other potentially climate-altering activities, such as deforestation.

ash. Impurities consisting of silica, iron, alumina, and other noncombustible matter that are contained in coal. Ash increases the weight of coal, adds to the cost of handling, and can affect its burning characteristics. Ash content is measured as a percent by weight of coal on an *as received* or a *dry* (moisture-free, usually part of a laboratory analysis) basis.

asset. An economic resource, tangible or intangible, that is expected to provide benefits to a business.

auxiliary generator. A generator at the electric plant site that provides power for the operation of the electrical generating equipment itself, including related demands such as plant lighting, during periods when the electric plant is not operating and power is unavailable from the grid. A black-start generator used to start main central station generators is considered to be an auxiliary generator.

auxiliary power units (APUs). Small gas turbines designed for auxiliary power of larger machines.

available but not needed capability. Net capability of main generating units that are operable but not considered necessary to carry load and cannot be connected to load within 30 minutes.

available capacity. Installed capacity less generation out of service.

average revenue (per kilowatt-hour). The average revenue per kilowatt-hour of electricity sold by sector (residential, commercial, industrial, or other) and geographic area (state, census division, and national) is calculated by dividing the total monthly revenue by the corresponding total monthly sales for each sector and geographic area.

avoided cost. The incremental cost of electric power to an electric utility that the utility would generate itself or purchase from another source if it did not purchase the power from qualifying facilities (QFs). FERC regulations implemented PURPA, which requires that utilities purchase electricity from QFs at a price at or below their avoided costs.

B

backup fuel. In a central heat pump system, the fuel used in the furnace that takes over the space heating when the outdoor temperature drops below that which is feasible to operate a heat pump.

backup generator. A generator that is used only for test purposes, or in the event of an emergency, such as a shortage of power needed to meet customer load requirements.

backup power. Electric energy supplied by a utility to replace power and energy lost during an unscheduled equipment outage.

base bill. A charge calculated by taking the rate from the appropriate electric rate schedule and applying it to the level of consumption.

base load. The minimum amount of electric power delivered or required over a given period of time at a steady rate.

base load capacity. The generating equipment normally operated to serve loads on an around-the-clock basis.

base load plant. A plant, usually housing high-efficiency steam-electric units, which is normally operated to take all or part of the minimum load of a system, and which consequently produces electricity at an essentially constant rate and runs continuously. These units are operated to maximize system mechanical and thermal efficiency and minimize system operating costs.

base rate. A fixed kilowatt hour charge for electricity consumed that is independent of other charges and/or adjustments.

benzene (C_6H_6). An aromatic hydrocarbon present in small proportion in some crude oil and made commercially from petroleum by the catalytic reforming of naphthenes in petroleum naphtha. Also made from coal in the manufacture of coke. Used as a solvent in the manufacture of detergents, synthetic fibers, petrochemicals, and as a component of high-octane gasoline.

bi-fuel vehicle. A motor vehicle that operates on two different fuels, but not on a mixture of the fuels. Each fuel is stored in a separate tank.

bilateral agreement. A written statement signed by two parties that specifies the terms for exchanging energy.

bilateral electricity contract. A direct contract between an electric power producer and either a user or broker outside of a centralized power pool or power exchange.

billing period. The time between meter readings. It does not refer to the time when the bill was sent or when the payment was to have been received. In some cases, the billing period is the same as the billing cycle that corresponds closely (within several days) to meter-reading dates. For fuel oil and LPG, the billing period is the number of days between fuel deliveries.

biofuels. Liquid fuels and blending components produced from biomass (plant) feedstocks, used primarily for transportation.

biomass. Organic nonfossil material of biological origin constituting a renewable energy source.

biomass gas. A medium BTU gas containing methane and carbon dioxide, resulting from the action of microorganisms on organic materials, such as a landfill.

biomass resources. Any plant-derived organic matter available on a renewable basis, including dedicated energy crops and trees, agricultural food and feed crops, agricultural crop wastes and residues, wood wastes and residues, aquatic plants, animal wastes, municipal wastes, and other waste materials. Handling technologies, collection logistics and infrastructure are all important aspects of the biomass resources supply chain.

biomass technologies. Convert renewable biomass fuels into electricity (and heat) using modern boilers, gasifiers, turbines, generators, and fuel cells.

bituminous coal. A dense coal, usually black, sometimes dark brown, often with well-defined bands of bright and dull material, used primarily as fuel in steam-electric power generation, with substantial quantities also used for heat and power applications in manufacturing and to make coke. Bituminous coal is the most abundant coal in active U.S. mining regions. Its moisture content usually is less than 20%. The heat content of bituminous coal ranges from 21 to 30 million BTU per ton on a moist, mineral-matter-free basis. The heat content of bituminous coal consumed in the United States averages 24 million BTU per ton, on the as-received basis (i.e., containing both inherent moisture and mineral matter).

block-rate structure. An electric rate schedule with a provision for charging a different unit cost for various increasing blocks of demand for energy. A reduced rate may be charged on succeeding blocks.

boiler. A device for generating steam for power, processing, or heating purposes; or hot water for heating purposes or hot water supply. Heat from an external combustion source is transmitted to a fluid contained within the tubes found in the boiler shell. This fluid is delivered to an end use at a desired pressure, temperature, and quality.

boiler fuel. An energy source to produce heat that is transferred to the boiler vessel in order to generate steam or hot water. Fossil fuel is the primary energy source used to produce heat for boilers.

boiling-water reactor (BWR). A light-water reactor in which water, used as both coolant and moderator, is allowed to boil in the core. The resulting steam can be used directly to drive a turbine.

borderline customer. A customer located in the service area of one utility, but supplied by a neighboring utility through an arrangement between the utilities.

bottoming cycle. A waste-heat recovery boiler recaptures the unused energy and uses it to produce steam to drive a steam turbine generator to produce electricity.

Brayton cycle. Gas turbines are described thermodynamically by the Brayton cycle, in which air is compressed isentropically, combustion occurs at constant pressure, and expansion over the turbine occurs isentropically back to the starting pressure.

British thermal unit (BTU). A standard unit for measuring the quantity of heat energy equal to the quantity of heat required to raise the temperature of 1 pound of water by 1°F.

BTU conversion. BTU conversion factors for site energy are as follows:

Electricity	3,412 BTU/kWh
Natural gas	1,031 BTU/cf
Fuel oil No. 1	135,000 BTU/gal
Kerosene	135,000 BTU/gal
Fuel oil No. 2	138,690 BTU/gal
LPG (propane)	91,330 BTU/gal
Wood	20 million BTU/cord

bulk power transactions. The wholesale sale, purchase, and interchange of electricity among electric utilities. Bulk power transactions are used by electric utilities for many different aspects of electric utility operations, from maintaining load to reducing costs.

bundled utility service. A means of operation whereby energy, transmission, and distribution services, as well as ancillary and retail services, are provided by one entity.

bus. An electrical conductor that serves as a common connection for two or more electrical circuits.

busbar cost. The traditional measure for the total cost of generating electricity. Busbar cost refers to the cost of generated electricity before it enters the utility's system. The busbar is the conductor that takes electricity from the plant to the switchyard.

butylene (C_4H_8). An olefinic hydrocarbon recovered from refinery processes.

C

California Power Exchange. A state-chartered, nonprofit corporation that provides day-ahead and hour-ahead markets for energy and ancillary services in accordance with the power exchange tariff. The power exchange is a scheduling coordinator and is independent of both the ISO and all other market participants.

capability. The maximum load that a generating unit, generating stations, or other electrical apparatus can carry under specified conditions for a given period of time without exceeding approved limits of temperature and stress.

capacity. The amount of electric power delivered or required for which a generator, turbine, transformer, transmission circuit, station, or system is rated by the manufacturer.

capacity charge. An element in a two-part pricing method used in capacity transactions (energy charge is the other element). The capacity charge, sometimes called *demand charge*, is assessed on the amount of capacity being purchased.

capacity factor. The ratio of the electrical energy produced by a generating unit for the period of time considered to the electrical energy that could have been produced at continuous full power operation during the same period.

capital. The line items on the right side of a balance sheet that include debt, preferred stock, and common equity. A net increase in assets must be financed by an increase in one or more forms of capital.

carbon dioxide (CO_2). A colorless, odorless, nonpoisonous gas that is a normal part of Earth's atmosphere. Carbon dioxide is a product of fossil-fuel combustion as well as other processes. It is considered a greenhouse gas as it traps heat (infrared energy) radiated by the Earth into the atmosphere and thereby contributes to the potential for global warming. The global warming potential (GWP) of other greenhouse gases is measured in relation to that of carbon dioxide, which by international scientific convention is assigned a value of one (1).

carbon dioxide equivalent. The amount of CO_2 by weight emitted into the atmosphere that would produce the same estimated radiative forcing as a given weight of another radiatively active gas. Carbon dioxide equivalents are computed by multiplying the weight of the gas being measured (for example, methane) by its estimated global warming potential (which is 21 for methane). *Carbon equivalent units* are defined as carbon dioxide equivalents multiplied by the carbon content of CO_2 (i.e., 12/44).

chlorofluorocarbon (CFC). Any of various compounds consisting of carbon, hydrogen, chlorine, and fluorine used as refrigerants. CFCs are now thought to be harmful to the Earth's atmosphere.

circuit. Refers to a conductor or a system of conductors through which electric current flows.

class rate schedule. An electric rate schedule applicable to one or more specified classes of service, groups of businesses, or customer uses.

clean coal technology (CCT). CCT technologies are the products of research and development conducted over the past 20 years. The result has been more than 20 new, lower-cost, more efficient and environmentally compatible technologies for electric utilities, as well as for other industries. The original CCT program, which began in 1986, focused on commercializing processes that helped reduce sulfur dioxide and nitrogen oxide emissions and demonstrating more efficient and environmentally friendly alternatives to traditional pulverized coal boilers. New programs, such as the CCPI, are extensions of the original CCT program. These new programs are finding solutions for reducing trace emissions in mercury, reducing or eliminating CO_2 emissions, and increasing fuel efficiencies. Over the longer terms, research in CCT will be directed toward developing coal-based hydrogen fuels. If coupled with sequestration, this will allow greater use of coal with zero emissions.

coal. A black or brownish-black solid combustible substance formed by the partial decomposition of vegetable matter without access to air. The rank of coal, which includes anthracite, bituminous, subbituminous, and lignite, is based on fixed carbon, volatile matter, and heating value. Coal rank indicates the progressive alteration from lignite to anthracite. Lignite contains approximately 9 to 17 million BTU/t. The contents of subbituminous and bituminous coal range from 16 to 24 million BTU/t, and from 19 to 30 million BTU/t, respectively. Anthracite contains approximately 22 to 28 million BTU/t.

coal bed degasification. The removal of methane or coal bed gas from a coal mine before or during mining.

coal gasification. The process of converting coal into gas. The basic process involves crushing coal to a powder, which is then heated in the presence of steam and oxygen to produce a gas. The gas is then refined to reduce sulfur and other impurities. The gas can be used as a fuel or processed further and concentrated into chemical or liquid fuel.

cofiring. The process of burning natural gas in conjunction with another fuel to reduce air pollutants.

cogeneration. The production of electrical energy and another form of useful energy (such as heat or steam) through the sequential use of energy.

cogeneration system. A system using a common energy source to produce both electricity and steam for other uses, resulting in increased fuel efficiency.

cogenerator. A generating facility that produces electricity and another form of useful thermal energy (such as heat or steam) used for industrial, commercial, heating, or cooling purposes. To receive status as a QF under PURPA, the facility must produce electric energy and "another form of useful thermal energy through the sequential use of energy" and meet certain ownership, operating, and efficiency criteria established by the FERC. (See the Code of Federal Regulations, Title 18, Part 292, www.gpoaccess.gov/cfr/index.html.)

coincidental demand. The sum of two or more demands occurring in the same time interval.

coincidental peak load. The sum of two or more peak loads occurring in the same time interval.

combined cycle. An electric generating technology in which electricity is produced from otherwise lost waste heat exiting from one or more gas (combustion) turbines. The exiting heat is routed to a conventional boiler or to a heat recovery steam generator for utilization by a steam turbine in the production of electricity. This process increases the efficiency of the electric generating unit.

combined heat and power (CHP) plant. A plant designed to produce both heat and electricity from a single heat source. (This term is being used in place of the term *cogenerator* that was used by the EIA in the past. CHP better describes the facilities because some of the plants included do not produce heat and power in a sequential fashion and, as a result, do not meet the legal definition of cogeneration specified in the PURPA.)

combined hydroelectric plant. A hydroelectric plant that uses both pumped water and natural stream flow for the production of power.

combined pumped-storage plant. A pumped-storage hydroelectric power plant that uses both pumped water and natural stream flow to produce electricity.

combined-cycle unit. An electric generating unit that consists of one or more combustion turbines and one or more boilers with a portion of the required energy input to the boiler(s) provided by the exhaust gas of the combustion turbine(s).

commercial building. A building with more than 50% of its floor space used for commercial activities. Commercial buildings include, but are not limited to, stores, offices, schools, churches, gymnasiums, libraries, museums, hospitals, clinics, warehouses, and jails. Government buildings are included except for buildings on military bases or reservations.

commercial facility. An economic unit that is owned or operated by one person or organization and that occupies two or more commercial buildings at a single location. A university or a large hospital complex is an example of a commercial multibuilding facility.

commercial sector. An energy-consuming sector that consists of service-providing facilities and equipment of: businesses; federal, state, and local governments; and other private and public organizations, such as religious, social, or fraternal groups. The commercial sector includes institutional living quarters. It also includes sewage treatment facilities. Common uses of energy associated with this sector include space heating, water heating, air conditioning, lighting, refrigeration, cooking, and running a wide variety of other equipment. This sector also includes generators that produce electricity and/or useful thermal output primarily to support the activities of the above-mentioned commercial establishments.

commingling. The mixing of one utility's generated supply of electric energy with another utility's generated supply within a transmission system.

competitive bidding. A procedure that utilities in many states use to select suppliers of new electric capacity and energy. Under competitive bidding, an electric utility solicits bids from prospective power generators to meet current or future power demands. When offers from IPPs began exceeding utility needs in the mid-1980s, utilities and state regulators began using competitive bidding systems to select from among numerous supply alternatives.

competitive transition charge. A required charge levied on each customer of the distribution utility, including those who are served under contracts with nonutility suppliers, for recovery of the utility's stranded costs that develop because of competition.

concentrating solar power. As a solar technology, concentrating solar power uses reflective materials, such as mirrors, to concentrate solar energy. This concentrated heat energy is then converted into electricity. Power plants can use a concentrating solar system. The sunlight is collected and focused with mirrors to create a high-intensity power source. This heat source produces steam or mechanical power to run a generator that creates electricity.

congestion. A condition that occurs when insufficient transfer capacity is available to implement all of the preferred schedules for electricity transmission simultaneously.

connected load. The sum of the continuous ratings or the capacities for a system, part of a system, or a customer's electric power consuming apparatus.

connection. Term referring to the physical connection of transmission lines, transformers, switch gear, etc., between two electric systems permitting the transfer of electric energy in one or both directions.

construction work in progress (CWIP). CWIP is the balance shown on a utility's balance sheet for construction work not yet completed but in process. This balance line item may or may not be included in the rate base.

consumption. The amounts of fuel used for gross generation, providing standby service, start-up, and/or flam stabilization.

contract price. Price of fuels marketed on a contract basis covering a period of one or more years. Contract prices reflect market conditions at the time the contract was negotiated and therefore remain constant throughout the life of the contract or are adjusted through escalation clauses. Generally, contract prices do not fluctuate widely.

contract receipts. Purchases based on a negotiated agreement that generally covers one or more years.

control area. An electric power system or combination of electric power systems. A common automatic control scheme is applied to the systems. This is done in order to match the power output of the generators within the electric power system and capacity and energy purchased from entities outside the electric power system with the load in the electric power system. The control area maintains scheduled interchange with other control areas and maintains the frequency of the electric power system within reasonable limits. It also provides sufficient generating capacity to maintain operating reserves.

cooperative electric utility. Established to be owned by and operated for the benefit of those using its service. The utility company will generate, transmit, and/or distribute supplies of electric energy to a specified area not being serviced by another utility. Such ventures are generally exempt from federal income tax laws. Most electric cooperatives have been initially financed by the RUS (formerly the REA), U.S. Department of Agriculture.

cross-subsidization. The transfer of assets or services from the regulated portion of an electric utility to its unregulated affiliates to produce an unfair competitive advantage. This can also refer to one rate class, such as industrial customers, subsidizing the rates of another class, such as residential customers.

current. A flow of electrical charge in a conductor is referred to as a current. The strength or rate of movement of the electricity is measured in amperes.

customer choice. The right of customers to purchase energy from a supplier other than their traditional supplier or from more than one seller in the retail market.

D

deforestation. The net removal of trees from forested land.

degasification system. The methods employed for removing methane from a coal seam that could not otherwise be removed by standard ventilation fans and thus would pose a substantial hazard to coal miners. These systems may be used prior to mining or during mining activities.

delivering party. The entity supplying the capacity and/or energy to be transmitted at point(s) of receipt.

demand. The rate at which electric energy is delivered to or by a system, part of a system, or piece of equipment, at a given instant or averaged over any designated period of time (see **peak demand**).

demand charge. That portion of the consumer's bill for electric service based on the consumer's maximum electric capacity usage and calculated based on the billing demand charges under the applicable rate schedule.

demand charge credit. Compensation received by the buyer when the delivery terms of the contract cannot be met by the seller.

demand indicator. A measure of the number of energy-consuming units, or the amount of service or output, for which energy inputs are required.

demand interval. The time period during which flow of electricity is measured (usually in 15-, 30-, or 60-minute increments).

demand metered. Having a meter to measure peak demand (in addition to total consumption) during a billing period. Demand is not usually metered for other energy sources.

demand-side management (DSM). The planning, implementation, and monitoring of utility activities designed to encourage consumers to modify patterns of electricity usage, including the timing and level of electricity demand. It refers to only energy and load-shape modifying activities that are undertaken in response to utility-administered programs. It does not refer to energy and load-shaped changes arising from the normal operation of the marketplace or from government-mandated energy-efficiency standards. DSM covers the complete range of load-shape objectives, including strategic conservation and load management, as well as strategic load growth.

demand-side management costs. The costs incurred by the utility to achieve the capacity and energy savings from a DSM program. Costs incurred by customers or third parties are to be excluded. The costs are to be reported in thousands of dollars (nominal) in the year in which they are incurred, regardless of when the savings occur. The utility costs are all the annual expenses (labor, administrative, equipment, incentives, marketing, monitoring and evaluation, and other costs incurred by the utility for operation of the DSM program), regardless of whether the costs are expensed or capitalized. Lump-sum capital costs (typically accrued over several years prior to start-up) are not to be reported. Program costs associated with strategic load growth activities are also to be excluded.

Department of Energy (DOE). Manages programs of research, development, and commercialization for various energy technologies and associated environmental, regulatory, and defense programs. DOE promulgates energy policies and acts as a principal adviser to the president of the United States on energy matters.

deregulation. The elimination of some or all regulations from a previously regulated industry or sector of an industry.

designated agent. Any entity that performs actions or functions on behalf of the transmission provider, an eligible customer, or the transmission customer required under the tariff.

direct access. The ability of a retail customer to purchase electricity or other energy sources directly from a supplier other than its traditional supplier.

direct control load management. The magnitude of customer demand that can be interrupted at the time of the seasonal peak load by direct control of the system operator by interrupting power supply to individual appliances or equipment on customer premises. This type of control usually reduces the demand of residential customers.

direct current (DC). An electric current of constant direction, having a magnitude that does not vary or varies only slightly as in a battery. (Compare with **alternating current**.)

direct electricity demand control. The utility installs a radio-controlled device on the HVAC equipment. During periods of particularly heavy use of electricity, the utility will send a radio signal to the building in its service territory with this device and turn off the HVAC for a certain period.

direct load control. This DSM category represents the consumer load that can be interrupted at the time of annual peak load by direct control of the utility system operator. Direct load control does not include interruptible load. This type of control usually involves residential consumers.

direct use. Use of electricity that is: 1) self-generated; 2) produced by either the same entity that consumes the power or an affiliate; and 3) used in direct support of a service or industrial process located within the same facility or group of facilities that house the generating equipment. Direct use is exclusive of station use.

direct utility cost. A utility cost that is identified with one of the DSM program categories (e.g., energy efficiency or load management).

dispatchability. The ability of a generating unit to increase or decrease generation, or to be brought on line or shut down at the request of a utility's system operator.

dispatching. The operating control of an integrated electric system involving operations such as: 1) the assignment of load to specific generating stations and other sources of supply to effect the most economical supply as the total or the significant area loads rise or fall; 2) the control of operations and maintenance of high-voltage lines, substations, and equipment; 3) the operation of principal tie-lines and switching; 4) the scheduling of energy transactions with connecting electric utilities.

distillate fuel oil. A general classification for one of the petroleum fractions produced in conventional distillation operations. It is used primarily for space heating, on-and-off-highway diesel engine fuel (including railroad engine fuel and fuel for agricultural machinery), and electric power generation. Included are fuels oils No. 1, No. 2, and No. 4, and diesel fuels No. 1, No. 2, and No. 4.

distribution. The delivery of energy to retail customers.

distribution system. The portion of the transmission and facilities of an electric system that is dedicated to delivering electric energy to an end user.

district heat. Steam or hot water from an outside source used as an energy source in a building The steam or hot water is produced in a central plant and piped into the building. The district heat may be purchased from a utility or provided by a physical plant in a separate building that is part of the same facility (for example, a hospital complex or university).

diversity exchange. An exchange of capacity or energy, or both, between systems, whose peak loads occur at different times.

divestiture. The stripping off of one utility function from the others by selling (spinning off) or in some other way changing the ownership of the assets related to that function. Stripping off is most commonly associated with spinning off generation assets so they are no longer owned by the shareholders who own the transmission and distribution assets.

dual-fired unit. A generating unit that can produce electricity using two or more input fuels. In some of these units, only the primary fuel can be used continuously; the alternate fuel(s) can be used only as a start-up fuel or in emergencies.

dynamic reactive power sources. Also referred to as fast-switched reactive power sources, these sources of power include generators, synchronous condensers, and static volt-ampere reactive (VAR) devices. Generally, dynamic reactive power sources are considered to be sources that can respond instantaneously to voltage changes. All of these reactive power sources provide immediate and automatic support for a system whenever a systems contingency occurs. (Compare with **static reactive power devices.**)

E

Edison Electric Institute (EEI). The membership association of the U.S. electric IOUs and industry affiliates worldwide. Organized in 1933 and incorporated in 1970, EEI provides a principal forum where electric utility professionals exchange information on developments in their business and operate as a liaison between the industry and the federal government. EEI's officers act as spokespersons for IOUs on subjects of national interest. EEI's stated basic objective reads, "In its leadership role, EEI provides advocacy, authoritative analysis, and critical industry data to its members, Congress, government agencies, the financial community, and other opinion-leader audiences." EEI compiles factual information, data, and statistics relating to the electric industry, and makes them available to member companies, the public, and government representatives. Information accessed November 30, 2005. See www.eei.org/about_eei/index.htm for more information.

electric capacity. The ability of a power plant to produce a given output of electric energy at an instant in time. Capacity is measured in kilowatts or megawatts.

electric expenses. The cost of labor, material, and expenses incurred in operating a facility's prime movers, generators, auxiliary apparatus, switching gear, and other electric equipment for each of the points where electricity enters the transmission or distribution grid.

electric generation industry. Stationary and mobile generating units that are connected to the electric power grid and can generate electricity. The electric generation industry includes the *electric power sector* (utility generators and IPPs) and industrial and commercial power generators, including combined-heat-and-power producers, but excludes units at single-family dwellings.

electric generator. A facility that produces only electricity, commonly expressed in kilowatt-hours (kWh) or megawatt-hours (MWh). Electric generators include electric utilities and IPPs.

electric hybrid vehicle. An electric vehicle that either: 1) operates solely on electricity, but contains an internal combustion motor that generates additional electricity (series hybrid); or 2) contains an electric system and an internal combustion system and is capable of operating on either system (parallel hybrid).

electric industry reregulation. The design and implementation of regulatory practices to be applied to the remaining traditional utilities after the electric power industry has been restructured. Reregulation applies to those entities that continue to exhibit characteristics of a natural monopoly. Reregulation could employ the same or different regulatory practices as those used before restructuring.

electric industry restructuring. The process of replacing a monopolistic system of electric utility suppliers with competing sellers, allowing individual retail customers to choose their supplier but still receive delivery over the power lines of the local utility. It includes the reconfiguration of vertically-integrated electric utilities.

electric operating expenses. Summation of electric operation-related expenses, such as operation expenses, maintenance expenses, depreciation expenses, amortization, taxes other than income taxes, federal income taxes, other income taxes, provision for deferred income taxes, provision for deferred income-credit, and investment tax credit adjustment.

electric plant (physical). A facility containing prime movers, electric generators, and auxiliary equipment for converting mechanical, chemical, and/or fission energy into electric energy.

electric power. The rate at which electric energy is transferred. Electric power is measured by capacity and is commonly expressed in megawatts (MW).

electric power grid. A system of synchronized power providers and consumers connected by transmission and distribution lines and operated by one or more control centers. In the continental United States, the electric power grid consists of three systems: the Eastern

Interconnect, the Western Interconnect, and the Texas Interconnect. In Alaska and Hawaii, several systems encompass areas smaller than the state (e.g., the interconnect serving Anchorage, Fairbanks, and the Kenai Peninsula; and individual islands).

electric power plant. A station containing prime movers, electric generators, and auxiliary equipment for converting mechanical, chemical, and/or fission energy into electric energy.

electric power sector. An energy-consuming sector that consists of electricity only and CHP plants whose primary business is to sell electricity, or electricity and heat, to the public (i.e., North American Industry Classification System 22 plants). See also **combined heat and power (CHP) plant**.

electric power system. An individual electric power entity: a company; an electric cooperative; a public electric supply corporation as the TVA; a similar federal department or agency, such as the BPA; the Bureau of Reclamation or the Corps of Engineers; a municipally owned electric department offering service to the public; or an electric public utility district (PUD); also a jointly owned electric supply project such as the Keystone.

electric rate. The price set for a specified amount and type of electricity by class of service in an electric rate schedule or sales contract.

electric rate schedule. A statement of the electric rate and the terms and conditions governing its application, including attendant contract terms and conditions that have been accepted by a regulatory body with appropriate oversight authority.

electric system reliability. The degree to which the performance of the elements of the electrical system results in power being delivered to consumers within accepted standards and in the amount desired. Reliability encompasses two concepts: adequacy and security. *Adequacy* implies that there are sufficient generation and transmission resources installed and available to meet projected electrical demand plus reserves for contingencies. *Security* implies that the system will remain intact operationally (i.e., will have sufficient available operating capacity) even after outages or other equipment failure. The degree of reliability may be measured by the frequency, duration, and magnitude of adverse effects on consumer service.

electric utility. A corporation, person, agency, authority, or other legal entity or instrumentality aligned with distribution facilities for delivery of electric energy for use primarily by the public. Included are electric IOUs, municipal and state utilities, federal electric utilities, and rural electric cooperatives. A few entities that are tariff based and corporately aligned with companies that own distribution facilities are also included.

electric utility divestiture. The separation of one electric utility function from others through the selling of the management and ownership of the assets related to that function. It is most commonly associated with selling generation assets so they are no longer owned or controlled by the shareholders that own the company's transmission and distribution assets.

electric utility restructuring. The introduction of competition into at least the generation phase of electricity production, with a corresponding decrease in regulatory control.

electric utility sector. The electric utility sector consists of privately and publicly owned establishments that generate, transmit, distribute, or sell electricity primarily for use by the public and that meet the definition of an electric utility. Nonutility power producers are not included in the electric sector.

electrical current. If two equally and oppositely charged bodies are connected by a metallic conductor such as a wire, the charges neutralize each other. This neutralization is accomplished by means of a flow of electrons through the conductor from the negatively charged body to the positively charged one. In any continuous system of conductors, electrons will flow from the point of lowest potential to the point of highest potential. A system of this kind is called an electric current. The current flowing in a circuit is described as direct current, or DC, if it flows continuously in one direction, and as alternating current, or AC, if it flows alternately in either direction.

electricity. A class of physical phenomena resulting from the existence of charge and from the interaction of charges. When a charge is stationary or static, it produces forces on objects in regions where it is present, and when it is in motion, it produces magnetic effects. Electric and magnetic effects are caused by the relative position and movement of positively and negatively charged particles of matter. So far as electrical effects are concerned, these particles are neutral, positive, or negative. Electricity is concerned with the positively charged particles, such as protons, that repel one another, and the negatively charged particles, such as electrons, that also repel one another. Negative and positive particles, however, attract each other. This behavior may be summarized as follows: Like charges repel, and unlike charges attract.

electricity broker. An entity that arranges the sale and purchase of electric energy, the transmission of electricity, and/or other related services between buyers and sellers but does not take title to any of the power sold.

end user. A firm or individual that purchases products for its own consumption and not for resale (i.e., an ultimate consumer).

energy. The capacity for doing work as measured by the capability of doing work (potential energy) or the conversion of this capability to motion (kinetic energy). Energy has several forms, some of which are easily convertible and can be changed to another form useful for work. Most of the world's convertible energy comes from fossil fuels that are burned to produce heat that is then used as a transfer medium to mechanical or other means in order to accomplish tasks. Electrical energy is usually measured in kilowatt-hours, while heat energy is usually measured in British thermal units (BTU).

energy broker system. Introduced into Florida by the Public Service Commission, the energy broker system is a system for exchanging information that allows utilities to efficiently exchange hourly quotations of prices at which each is willing to buy and sell electric energy. For the broker system to operate, utility systems must have in place bilateral agreements between all potential parties, must have transmission arrangements between all potential parties, and must have transmission arrangements that allow the exchanges to take place.

energy effects. The changes in aggregate electricity use (measured in megawatt-hours) for customers who participate in a utility DSM program. Energy effects should represent changes at the customer meter and reflect only activities that are undertaken specifically in response to utility-administered programs. This includes those activities implemented by the third parties under contract to the utility.

Energy Information Administration (EIA). An independent agency within the U.S. DOE that develops surveys, collects energy data, and does analytical and modeling analyses of energy issues. The agency must satisfy the requests of Congress, other elements within the DOE, FERC, the executive branch, its own independent needs, and assist the general public, or other interest groups, without taking a policy position.

energy management and control system (EMCS). An energy conservation feature that uses mini/microcomputers, instrumentation, control equipment, and software to manage a building's use of energy for heating, ventilation, air conditioning, lighting, and/or business-related processes. These systems can also manage fire control, safety, and security. Not included as EMCS are time-clock thermostats.

energy marketer. An entity, regulated by the FERC, which arranges bulk power transactions for end users. An energy marketer's main function is to determine the best overall fuel choice(s) for customers and then to deliver that fuel to the customer. Compare with **power marketer.**

Energy Policy Act of 1992 (EPAct of 1992). Helped to create a more competitive U.S. electric power marketplace by removing barriers to competition. By doing so, this act allowed for a broad spectrum of independent energy producers to compete in wholesale electric power markets. It also made significant changes in the way power transmission grids are regulated. Specifically, the law gave FERC the authority to order electric utilities to provide access to their transmission facilities to other power suppliers.

Energy Policy Act of 2005 (EPAct of 2005). Signed into law August 8, 2005, EPAct 2005 is the first effort of the United States government to address U.S. energy policy since the Energy Policy Act of 1992. The 1,724-page act (H.R. 6) encourages energy efficiency and conservation, promotes alternative and renewable energy sources, reduces our dependence on foreign sources of energy, increases domestic production, modernizes the electricity grid, and encourages the expansion of nuclear energy. In the area of solar and energy efficiency measures, EPAct 2005 created a number of tax credit opportunities. These include tax credits for residential solar photovoltaic and hot water heating systems, tax deductions for highly efficient commercial buildings, tax credits for high efficient new homes, tax credits for improvements to existing homes including high efficiency air conditioners and equipment, tax credits for residential fuel cell systems, and tax credits for fuel cell and microturbines used in businesses.

energy service provider (ESP). An energy entity that provides service to a retail or end-use customer.

Environmental Protection Agency (EPA). A federal agency that administers federal environmental policies, enforces environmental laws and regulations, performs research, and provides information on environmental subjects. EPA acts as the chief adviser to the president on U.S. environmental policy and issues.

exempt wholesale generator (EWG). Wholesale generators created under the 1992 Energy Policy Act that are exempt from certain financial and legal restrictions stipulated in the PUHCA.

F

facilities charge. An amount to be paid by the customer in a lump sum or periodically as reimbursement for facilities furnished. The charge may include operation and maintenance as well as fixed costs.

fast breeder reactor (FBR). A breeder reactor that requires high-speed neutrons to produce fissionable material. A fast neutron reactor, commonly called simply *fast reactor*, is a nuclear reactor design that uses no moderator but instead relies on fast neutrons to sustain its chain reaction. Achieving this requires high-grade fuel such as enriched uranium or plutonium, but once this has been provided for the initial start-up, the reactor produces its own fuel and a surplus that can then be used to start other FBRs, hence the concept of a *breeder*.

federal electric utilities. An electric utility classification applying to those utilities that are agencies of the federal government involved in the generation and/or transmission of electricity. Most of the electricity generated by federal electric utilities is sold at wholesale prices to local government-owned and cooperatively owned utilities and to IOUs. These government agencies are the Army Corps of Engineers and the Bureau of Reclamation, which generate electricity at federally owned hydroelectric projects. There are five power marketing agencies that sell this relatively low-cost power on a preferential basis to local government-owned and cooperatively owned utilities. Also, the TVA produces and transmits electricity in the Tennessee Valley region.

Federal Energy Regulatory Commission (FERC). The chief energy regulatory body of the U.S. government. FERC was given new powers when Congress passed the Energy Policy Act of 1992 (EPAct of 1992). Under the act, FERC became responsible for determining EWG status and was given the authority to order utilities to provide access to their power transmission systems to other electric generators. In addition, FERC certifies QFs as defined by PURPA, establishes and enforces rates for power sales and transmission services, issues licenses for hydroelectric projects, and regulates certain aspects of mergers and acquisitions of gas and electric utility companies. The commission also establishes and enforces rates related to the sale and transportation of oil and natural gas. With the passage of EPAct of 2005, the FERC's role has changed. Overall, the Act gave FERC many new responsibilities in the area of electric regulation. The Act gives FERC the duty of assuring the reliability of the bulk power system. FERC will exercise that duty by certifying an Electric Reliability Organization (ERO), reviewing reliability standards, approving standards that provide for reliable operation of the bulk power system, remanding those that do not, and working to improve reliability standards over time. The FERC will also ensure reliability standards are properly enforced, including the regional enforcement of those standards. FERC has been given the authority to prevent the accumulation and exercise of generation market power by granting the agency authority to review acquisitions and transfers of generation facilities. It also gives FERC significant penalty authority.

Federal Power Act (FPA). Enacted in 1920, and amended in 1935, the act consists of three parts. The first part incorporated the Federal Water Power Act administered by the former Federal Power Commission, whose activities were confined almost entirely to licensing non-federal hydroelectric projects. Parts II and III were added with the passage of the Public Utility Act. These parts extended the act's jurisdiction to include regulating the interstate transmission of electrical energy and rates for its sale as wholesale in interstate commerce. The FERC is now charged with the administration of this law.

Federal Power Commission (FPC). The predecessor agency of FERC. The FPC was created under the Federal Water Power Act on June 10, 1920. It was charged originally with regulating the electric power and natural gas industries. It was abolished on September 30, 1977, when the DOE was created. Its functions were divided between the DOE and FERC, an independent regulatory agency.

firm power. Power or power-producing capacity, intended to be available at all times during the period covered by a guaranteed commitment to deliver, even under adverse conditions.

fission. The process whereby an atomic nucleus of appropriate type, after capturing a neutron, splits into (generally) two nuclei of lighter elements, with the release of substantial amounts of energy and two or more neutrons.

flue gas desulfurization. Equipment used to remove sulfur oxides from the combustion gases of a boiler plant before discharge to the atmosphere. Also referred to as *scrubbers*. Chemicals such as lime are used as scrubbing media.

fluidized-bed combustion. A method of burning particulate fuel, such as coal, in which the amount of air required for combustion far exceeds that found in conventional burners. The fuel particles are continually fed into a bed of mineral ash in the proportions of 1 part fuel to 200 parts ash, while a flow of air passes up through the bed, causing it to act like a turbulent fluid.

fly ash. Particulate matter mainly from coal ash in which the particle diameter is less than 1×10^{-4} meter. This ash is removed from the flue gas using flue gas particulate collectors such as fabric filters and electrostatic precipitators.

forced outage. The shutdown of a generating unit, transmission line, or other facility for emergency reasons or a condition in which the generating equipment is unavailable for load due to unanticipated breakdown.

fossil fuel. An energy source formed in the Earth's crust from decayed organic material. The common fossil fuels are petroleum, coal, and natural gas.

fossil fuel steam-electric plant. An electricity generation plant in which the prime mover is a turbine rotated by high-pressure steam produced in a boiler by heat from burning fossil fuels.

fuel. Any material substance that can be consumed to supply heat or power. Included are petroleum, coal, and natural gas (the fossil fuels), and other consumable materials, such as uranium, biomass, and hydrogen.

fuel cell. Device capable of generating an electrical current by converting the chemical energy of a fuel directly into electrical energy. Fuel cells differ from conventional electrical cells in that the active materials such as fuel and oxygen are not contained within the cell but are supplied from outside. It does not contain an intermediate heat cycle, as do most other electrical generation techniques.

full forced outage. The net capability of main generating units that are unavailable for load for emergency reasons.

G

gas. A fuel burned under boilers and by internal combustion engines for electric generation. These include natural, manufactured, and waste gas.

gas turbine. A rotary engine that extracts energy from a flow of combustion gas. It has an upstream compressor coupled to a downstream turbine, and a combustion chamber in between. Gas turbine may also refer to just the turbine element. Energy is added to the gas stream in the combustor, where air is mixed with fuel and ignited. Combustion increases the temperature and volume of the gas flow. This is directed through a nozzle over the turbine's blades, spinning the turbine and powering the compressor. Energy is extracted in the form of shaft power, compressed air, or thrust, and used to power generators, aircraft, trains, and ships. Power plant gas turbines can range in size from truck-mounted mobile plants to enormous, complex systems. They can be highly efficient—up to 60%—when waste heat from the gas turbine is recovered by a conventional steam turbine in a combined cycle (see **combined-cycle units**). Their main advantage is that they can be turned on and off within minutes, supplying power during peak demand. Large turbines may produce hundreds of megawatts.

generating facility. An existing or planned location or site at which electricity is or will be produced.

generating station. A station that consists of electric generators and auxiliary equipment for converting mechanical, chemical, or nuclear energy into electric energy.

generating unit. Any combination of physically connected generators, reactors, boilers, combustion turbines, and other prime movers operated together to produce electric power.

generation. The process of producing electric energy or the amount of electric energy produced by transforming other forms of energy, commonly expressed in kilowatt-hours (kWh) or megawatt-hours (MWh).

generation company. An entity that owns or operates generating plants. The generation company may own the generation plants or interact with the short-term market on behalf of plant owners.

generator capacity. The maximum output, commonly expressed in megawatts (MW), that generating equipment can supply to system load, adjusted for ambient conditions.

generator nameplate capacity (installed). The maximum rated output of a generator, prime mover, or other electric power production equipment under specific conditions designated by the manufacturer. Installed generator nameplate capacity is commonly expressed in megawatts (MW) and is usually indicated on a nameplate physically attached to the generator.

geothermal energy. Hot water or steam extracted from geothermal reservoirs in the Earth's crust. Water or steam extracted from geothermal reservoirs can be used for geothermal heat pumps, water heating, or electricity generation.

geothermal plant. A plant in which the prime mover is a steam turbine. The turbine is driven either by steam produced from hot water or by natural steam that derives its energy from heat found in rocks or fluids at various depths beneath the surface of the Earth. The energy is extracted by drilling and/or pumping.

gigawatt (GW). This is 1 billion W or 1,000 MW.

gigawatt-electric (GWe). This is 1 billion W of electric capacity.

gigawatt-hour (GWh). This is 1 billion Wh.

grid. The network of high-voltage transmission lines along which electrical energy flows. In the United States, the grids include the Eastern Interconnect, the Texas Interconnect, and the WSCC.

gross generation. The total amount of electric energy produced by the generating units at a generating station or stations, measured at the generator terminals.

H

heat pump. Heating and/or cooling equipment that, during the heating season, draws heat into a building from outside and, during the cooling season, ejects heat from the building to the outside. Heat pumps are vapor-compression refrigeration systems whose indoor/outdoor coils are used reversibly as condensers or evaporators, depending on the need for heating or cooling.

heat pump (air source). An air-source heat pump is the most common type of heat pump. The heat pump absorbs heat from the outside air and transfers the heat to the space to be heated in the heating mode. In the cooling mode the heat pump absorbs heat from the space to be cooled and rejects the heat to the outside air. In the heating mode when the outside air approaches 32°F or less, air-source heat pumps lose efficiency and generally require a backup (resistance) heating system.

heat pump (geothermal). A heat pump in which the refrigerant exchanges heat (in a heat exchanger) with a fluid circulating through a connection medium (ground or ground water). The fluid is contained in a variety of loop (pipe) configurations depending on the temperature of the ground and the ground area available. Loops may be installed horizontally or vertically in the ground or submersed in a body of water.

heat pump efficiency. The efficiency of a heat pump, that is, the electrical energy to operate it, is directly related to temperatures between which it operates. Geothermal heat pumps are more efficient than conventional heat pumps or air conditioners that use the outdoor air since the temperature of the ground or ground water a few feet below the Earth's surface remains relatively constant throughout the year. It is more efficient in the winter to draw heat from the relatively warm ground than from the atmosphere, where the air temperature is much colder, and in summer transfer waste heat to the relatively cool ground than to hotter air. Geothermal heat pumps are generally more expensive ($2,000–$5,000) to install than outside air heat pumps. However, depending on the location, geothermal heat pumps can reduce energy consumption (operating cost) and correspondingly, emissions by more than 20% compared to high-efficiency outside air heat pumps. Geothermal heat pumps also use the waste heat from air-conditioning to provide free hot water heating in the summer.

heat rate. A measure of generating station thermal efficiency, generally expressed in BTU per net kilowatt-hour. It is computed by dividing the total BTU content of fuel burned for electric generation by the resulting net kilowatt-hour generation.

heating degree-days (HDD). A measure of how cold a location is over a period of time relative to a base temperature, most commonly specified as 65°F. The measure is computed for each day by subtracting the average of the day's high and low temperatures from the base temperature (65°F), with negative values set equal to zero. Each day's HDD is summed to create an HDD measure for a specified reference period. HDD is used in energy analysis as an indicator of space heating energy requirements or use.

hourly nonfirm transmission service. Point-to-point transmission that is scheduled and paid for on an as-available basis and is subject to interruption.

hybrid electric vehicles. Vehicles that combine the internal combustion engine of a conventional engine with the battery and electric motor of an electric vehicle. This results in twice the fuel economy of conventional vehicles. This combination offers the extended range and rapid refueling that consumers expect from a conventional vehicle, with a significant portion of the energy and environmental benefits of an electric vehicle.

hybrid transmission line. A double-circuit line that has one AC and one DC circuit. The AC circuit usually serves local loads along the line.

hydrocarbon. An organic chemical compound of hydrogen and carbon in the gaseous, liquid, or solid phase. The molecular structure of hydrocarbon compounds varies from the simplest (methane, a constituent of natural gas) to the very heavy and very complex.

hydroelectric power. The use of flowing water to produce electrical energy.

hydroelectric pumped storage. Process for storing electric energy in the form of water held in an upper reservoir. Water is pumped from a lower reservoir into an upper reservoir during off-peak periods. It is then released through hydroelectric turbines into the lower reservoir for generation during peak-demand periods for management of electrical system loads, or to help provide needed power during system emergencies.

hydrogen. A colorless, odorless, highly flammable gaseous element. It is the lightest of all gases and the most abundant element in the universe, occurring chiefly in combination with oxygen in water and also in acids, bases, alcohols, petroleum, and other hydrocarbons.

I

impedance. The opposition to power flow in an AC circuit. Also, any device that introduces such opposition in the form of resistance, reactance, or both. The impedance of a circuit or device is measured as the ratio of voltage to current, where a sinusoidal voltage and current of the same frequency are used for the measurement; it is measured in ohms.

inadvertent power exchange. An unintended power exchange among utilities that is either not previously agreed upon or in an amount different from the amount agreed upon.

incandescent lamp. A glass enclosure in which light is produced when a tungsten filament is electrically heated so that it glows. Much of the energy is converted into heat; therefore, this class of lamp is a relatively inefficient source of light. Included in this category are the familiar screw-in light bulbs, as well as somewhat more efficient lamps, such as tungsten halogen lamps, reflector or r-lamps, parabolic aluminized reflector (PAR) lamps, and ellipsoidal reflector (ER) lamps.

incentives DSM program assistance. This DSM program assistance offers monetary or nonmonetary awards to encourage consumers to buy energy-efficient equipment and to participate in programs designed to reduce energy usage. Examples of incentives are zero- or low-interest loans, rebates, and direct installation of low-cost measures, such as water heater wraps or duct work for distributing the cool air; the units condition air only in the room or areas where they are located.

incremental effects. The annual changes in energy use (measured in megawatt-hours) and peak load (measured in kilowatts) caused by new participants in existing DSM programs and all participants in new DSM programs during a given year. *Reported incremental effects* are annualized to indicate the program effects that would have occurred had these participants been initiated into the program on January 1 of the given year. Incremental effects are not simply the annual effects of a given year minus the annual effects of the prior year, since these net effects would fail to account for program attrition, equipment degradation, building demolition, and participant dropouts. Incremental effects are not a monthly disaggregate of the annual effects, but are the total year's effects of only the new participants and programs for that year.

incremental energy costs. The additional cost of producing and/or transmitting electric energy above some previously determined base cost.

independent power producer (IPP). A corporation, person, agency, authority, or other legal entity or instrumentality that owns or operates facilities for the generation of electricity for use primarily by the public, and that is not an electric utility.

independent system operator (ISO). Developed by the FERC, ISOs facilitate open access of the electric transmission system, manage dispatch of generation and congestion, and administer a regional spot market for energy, capacity, and ancillary services. The coordination of these three activities under a centralized operation serves to enhance the stability and reliability of the regional power grid. See also **regional transmission organization**.

indirect utility cost. A cost that may not be identified with any particular DSM program category. Indirect costs could be attributable to one of several accounts' cost categories. These may include administrative, marketing, monitoring and evaluation, and utility-earned incentives. Accounting costs that are known DMS program costs should not be reported under indirect utility cost. Rather, those costs should be reported as direct utility costs under the appropriate DSM program category.

industrial sector. Utilities may classify industrial service using Standard Industrial Classification (SIC) codes, or may base the service on demand or annual usage exceeding some specified time limit. The limit may be set by the utility based on the rate schedule of the utility. The industrial sector is most usually an umbrella term encompassing manufacturing, construction, mining, agriculture, fishing, and forestry establishments.

instantaneous peak demand. The maximum demand at the instant of greatest load.

integrated gasification combined-cycle technology. Coal, water, and oxygen are fed to gasifier, which produces syngas. This medium-BTU gas is cleaned (particulates and sulfur compounds removed) and is fed to a gas turbine. The hot exhaust of the gas turbine and heat recovered from the gasification process are routed through a heat-recovery generator to produce steam, which drives a steam turbine to produce electricity.

integrated resource planning (IRP). A process by which an electric utility plans for its future resource needs. Key characteristics of IRP include a long-term forecast of power needs, a comprehensive evaluation of all resource options, both supply- and demand-side, and public review of the process.

intensity. The amount of a quantity per unit floor space. This method adjusts either the amount of energy consumed or expenditures spent, for the effects of various building characteristics, such as size of the building, number of workers, or number of operating hours, to facilitate comparisons of energy across time, fuels, and buildings.

intensity per hour. Total consumption of a particular fuel(s) divided by the total floor space of buildings that use the fuel(s) divided by total annual hours of operation.

interconnected system. A system consisting of two or more individual power systems normally operating with connecting tie-lines.

interconnection. Two or more electric systems having a common transmission line that permits a flow of energy between them. The physical connection of the electric power transmission facilities allows for the sale or exchange of energy.

interdepartmental service. Includes amounts charged by the electric department at tariff or other specified rates for electricity it supplies to other utility departments.

intermediate load. Refers to the range from base load to a point between base load and peak load. This stage may be the mid-point, a percent of the peak load, or the load over a specified time period.

intermittent electric generator. An electric generating plant with output controlled by the natural variability of the energy resource rather than dispatched based on system requirements. Intermittent output usually results from the direct, nonstored conversion of naturally occurring energy fluxes such as solar energy, wind energy, or the energy of free-flowing rivers (that is, run-of-river hydroelectricity).

internal combustion plant. A plant in which the prime mover is an internal combustion engine. An internal combustion engine has one or more cylinders in which the process of combustion takes place, converting energy released from the rapid burning of a fuel-air mixture into mechanical energy. Diesel or gas-fired engines are the principal types used in electric plants. The plant is usually operated during periods of high demand for electricity.

interruptible load. This DSM category represents the consumer load that, in accordance with contractual arrangements, can be interrupted at the time of annual peak load by the action of the consumer at the direct request of the system operator. This type of control usually involves large-volume commercial and industrial consumers. Interruptible load does not include direct load control.

interruptible power. Power and usually the associated energy made available by one utility to another. This transaction is subject to curtailment or cessation of delivery by the supplier in accordance with a prior agreement with the other party or under specified conditions.

interruptible rate. A special electricity or natural gas arrangement under which, in return for lower rates, the customer must either reduce energy demand on short notice or allow the electric or natural gas utility to temporarily cut off the energy supply for the utility to maintain service for higher priority users. This interruption or reduction in demand typically occurs during periods of high demand for the energy (summer for electricity and winter for natural gas).

investor-owned electric utilities (IOUs). A class of utilities that is funded from private investors. IOUs are owned by millions of investors, either directly or indirectly through other investments, such as life insurance policies, retirement funds, and mutual funds. IOUs sell their power to several different classes of customers and at wholesale rates (for resale) to state and local government-owned utilities, public utility districts, and rural electric cooperatives.

J

joule (J). The meter-kilogram-second unit of work or energy, equal to the work done by a force of 1 Newton (N) when its point of application moves through a distance of 1 m in the direction of the force; equivalent to 107 ergs and 1 watt-second.

Joule's Law. The rate of heat production by a steady current in any part of an electrical circuit that is proportional to the resistance and to the square of the current, or, the internal energy of an ideal gas depends only on its temperature.

jurisdictional utilities. Utilities regulated by public laws.

K

Kaplan turbine. A type of turbine that has two blades with adjustable pitch. The turbine may have gates to control the angle of the fluid flow into the blades.

kilovolt-ampere (kVA). A unit of apparent power, equal to 1,000 VA; the mathematical product of the volts and amperes in an electrical circuit.

kilowatt (kW). This is 1,000 W.

kilowatt-electric (kWe). This is 1,000 W of electric capacity.

kilowatt-hour (kWh). A measure of electricity defined as a unit of work or energy, measured as 1 kW (1,000 W) of power expended for 1 hour. One kWh is equivalent to 3,412 BTU.

L

leading edge. In reference to a wind energy conversion system, the area of a turbine blade surface that first comes into contact with the wind.

levelized cost. The present value of the total cost of building and operating a generating plant over its economic life, converted to equal annual payments. Costs are levelized in real dollars (i.e., adjusted to remove the impact of inflation).

leverage ratio. A measure that indicates the financial ability to meet debt service requirements and increase the value of the investment to the stockholders. (i.e., the ratio of total debt to total assets).

light water reactor (LWR). A nuclear reactor that uses water as the primary coolant and moderator, with slightly enriched uranium as fuel.

lighting DSM program. A DSM program designed to promote efficient lighting systems in new construction or existing facilities. Lighting DSM programs can include: certain types of high-efficiency fluorescent fixtures including T-8 lamp technology, solid-state electronic ballasts, specular reflectors, compact fluorescent fixtures, LED and electroluminescent emergency exit signs, high-pressure sodium with switchable ballasts, compact metal halide, occupancy sensors, and daylighting controllers.

lignite. The lowest rank of coal, often referred to as brown coal, used almost exclusively as fuel for steam-electric power generation. It is brownish-black and has a high inherent moisture content, sometimes as high as 45%. The heat content of lignite ranges from 9 to 17 million BTU per ton on a moist, mineral-matter-free basis. The heat content of lignite consumed in the United States averages 13 million BTU per ton, on the as-received basis (i.e., containing both inherent moisture and mineral matter).

line loss. Electric energy lost because of the transmission of electricity. Much of the loss is thermal in nature.

liquid metal fast breeder reactor. A nuclear breeder reactor, cooled by molten sodium, in which fission is caused by fast neutrons.

load control program. A program in which the utility company offers a lower rate in return for having permission to turn off the air conditioner or water heater for short periods of time by remote control. This control allows the utility to reduce peak demand.

load curve. The relationship of power supplied to the time of occurrence. Illustrates the varying magnitude of the load during the period covered.

load diversity. The difference between the peak of coincident and noncoincident demands of two or more individual loads.

load factor. The ratio of the average load to peak load during a specified time interval.

load following. Regulation of the power output of electric generators within a prescribed area in response to changes in system frequency, tie-line loading, or the relation of these to each other, so as to maintain the scheduled system frequency and/or established interchange with other areas within predetermined limits.

load leveling. Any load control technique that dampens the cyclical daily load flows and increases base load generation. Peak load pricing and time-of-day charges are two techniques that electric utilities use to reduce peak load and to maximize efficient generation of electricity.

load shape. A method of describing peak load demand and the relationship of power supplied to the time of occurrence.

load shedding. Intentional action by a utility that results in the reduction of more than 100 megawatts (MW) of firm customer load for reasons of maintaining the continuity of service of the reporting entity's bulk electric power supply system. The routine use of load control equipment that reduces firm customer load is not considered to be a reportable action.

local distribution company (LDC). A legal entity engaged primarily in the retail sale and/or delivery of natural gas through a distribution system that includes main lines (that is, pipelines designed to carry large volumes of gas, usually located under roads or other major right-of-ways) and laterals (that is, pipelines of smaller diameter that connect the end user to the main line). Since the restructuring of the gas industry, the sale of gas and/or delivery arrangements may be handled by other agents, such as producers, brokers, and marketers that are referred to as *non-LDC*.

Low Income Home Energy Assistance Program (LIHEAP). The purpose of LIHEAP is to assist eligible households to meet the cost of heating or cooling in residential dwellings. The federal government provides the funds to the states that administer the program.

M

maintenance expenses. That portion of operating expenses consisting of labor, materials, and other direct and indirect expenses incurred for preserving the operating efficiency and/or physical condition of utility plants used for power production, transmission, and distribution of energy.

maintenance supervision and engineering expenses. The cost of labor and expenses incurred in the general supervision and direction of the maintenance of power generation stations. The supervision and engineering included consists of the pay and expenses of superintendents, engineers, clerks, other employees, and consultants engaged in supervising and directing the maintenance of each utility function. Direct supervision and engineering of specific activities, such as fuel handling, boiler room operations, generator operations, etc., are charged to the appropriate accounts.

major electric utility. A utility that, in the last three consecutive calendar years, had sales or transmission services exceeding one of the following: 1) 1 million MWh of total annual sales; 2) 100 MWh of annual sales for resale; 3) 500 MWh of annual gross interchange out; or 4) 500 MWh of wheeling (deliveries plus losses) for others.

major energy sources. Fuels or energy sources such as electricity, fuel oil, natural gas, district steam, district hot water, and district chilled water. District chilled water is not included in any totals for the sum of major energy sources or fuels; all other major fuels are included in these totals.

major fuels. Fuels or energy sources such as: electricity, fuel oil, liquefied petroleum gases, natural gas, district steam, district hot water, and district chilled water.

manufacturing sector. An energy-consuming subsector of the industrial sector that consists of all facilities and equipment engaged in the mechanical, physical, chemical, or electronic transformation of materials, substances, or components into new products. Assembly of component parts of products is included, except for that which is included in construction.

marginal cost. The change in cost associated with a unit change in quantity supplied or produced.

market clearing price. The price at which supply equals demand for the day-ahead or hour-ahead markets.

market price contract. A contract in which the price of a product (uranium, for example) is not specifically determined at the time the contract is signed but is based instead on the prevailing market price at the time of delivery. A market price contract may include a floor price, that is, a lower limit on the eventual settled price. The floor price and the method of price escalation generally are determined when the contract is signed. The contract may also include a price ceiling or a discount from the agreed-upon market price reference.

market-based pricing. Prices of electric power or other forms of energy determined in an open market system of supply and demand under which prices are set solely by agreement as to what buyers will pay and sellers will accept. Such prices could recover less or more than full costs, depending upon what the buyers and sellers see as their relevant opportunities and risks.

marketing cost. Expenses directly associated with the preparation and implantation of strategies designed to encourage participation in a DSM program or other type program.

maximum demand. The greatest of all demands of the load that has occurred within a specified period of time.

megawatt electric (MWe). This represents 1 million W of electric capacity.

megawatt-hour (MWh). This represents 1,000 kWh or 1 million Wh.

merchant facilities. High-risk, high-profit facilities that operate, at least partially, at the whims of the market, as opposed to those facilities that are constructed with close cooperation of municipalities and have significant amounts of waste supply guaranteed.

merger. A combining of companies or corporations into one, often by issuing stock of the controlling corporation to replace the greater part of that of the other.

metered data. End-use data obtained through the direct measurement of the total energy consumed for specific uses within the individual household. Individual appliances can be submetered by connecting the recording meters directly to individual appliances.

metered peak demand. The presence of a device to measure the maximum rate of electricity consumption per unit of time. This device allows electric utility companies to bill their customers for maximum consumption, as well as for total consumption.

methane. A colorless, flammable, odorless hydrocarbon gas (CH_4) that is the major component of natural gas. It is also an important source of hydrogen in various industrial processes. Methane is a greenhouse gas.

mill. A monetary cost and billing unit used by utilities; it is equal to 1/1,000 of the U.S. dollar (equivalent to 1/10 of 1¢).

modules. Photovoltaic cells or an assembly of cells into panels (modules) intended for and shipped for final consumption or to another organization for resale. When exported, incomplete modules and cells that are not encapsulated cells are also included. Modules used for space applications are not included.

municipal electric utility. A class of electric utility owned by the city or municipality in which it operates. Municipal electric utilities are financed through municipal bonds and are self-regulated.

N

National Association of Regulatory Utility Commissioners (NARUC). An affiliation of the public service commissioners to promote the uniform treatment of members of the railroad, public utilities, and public service commissions of the 50 states, the District of Columbia, the Commonwealth of Puerto Rico, and the territory of the Virgin Islands.

National Electric Light Association (NELA). An early organization that governed the activities of investor-owned electric utilities. NELA was the forerunner of the EEI.

National Rural Electric Cooperative Association (NRECA). A national organization dedicated to representing the interests of cooperative electric utilities and the consumers they serve. Members come from the 46 states that have an electric distribution cooperative.

native load customers. The wholesale and retail customers on whose behalf the transmission provider, by statute, franchise, regulatory requirements, or contract has undertaken an obligation to construct and operate the transmission provider's system to meet the electric reliability needs of such customers.

Natural Gas Policy Act of 1978 (NGPA). Signed into law on November 9, 1978, the NGPA is a framework for the regulation of most facets of the natural gas industry.

net electricity consumption. Consumption of electricity computed as generation, plus imports, minus exports, minus transmission and distribution losses.

net energy for load. Net generation of main generating units that are system-owned or system-operated, plus energy receipts minus energy deliveries.

net energy for system. The sum of energy an electric utility needs to satisfy its service areas, including full and partial requirements consumers.

net generation. The amount of gross generation less the electrical energy consumed at the generating station(s) for station service or auxiliaries. (Electricity required for pumping at pumped-storage plants is regarded as electricity for station service and is deducted from gross generation.)

net interstate flow of electricity. The difference between the sum of electricity sales and losses within a state and the total amount of electricity generated within that state. A positive number indicates that more electricity (including associated losses) came into the state than went out of the state during the year; conversely, a negative number indicates that more electricity (including associated losses) went out of the state than came into the state.

net summer capacity. The maximum output, commonly expressed in megawatts (MW), that generating equipment can supply to system load, as demonstrated by a multihour test, at the time of summer peak demand (period of May 1 through October 31). This output reflects a reduction in capacity due to electricity use for station service or auxiliaries.

net winter capacity. The maximum output, commonly expressed in megawatts (MW), that generating equipment can supply to system load, as demonstrated by a multihour test, at the time of peak winter demand (period of November 1 though April 30). This output reflects a reduction in capacity due to electricity use for station service or auxiliaries.

nitrogen oxides (NO_x). Compounds of nitrogen and oxygen produced by the burning of fossil fuels.

nitrous oxide (N_2O). A colorless gas, naturally occurring in the atmosphere. Nitrous oxide has a 100-year global warming potential of 310.

nominal price. The price paid for a product or service at the time of the transaction. Nominal prices are those that have not been adjusted to remove the effect of changes in the purchasing power of the dollar; they reflect buying power in the year in which the transaction occurred.

nonfirm power. Power or power-producing capacity supplied or available under a commitment having limited or no assured availability.

nonrenewable fuels. Fuels that cannot be easily made or renewed; typically oil, natural gas, and coal.

nonrequirements consumer. A wholesale consumer (unlike a full or partial requirements consumer) that purchases economic or coordination power to supplement its own or another system's energy needs.

nonspinning reserve. The generating capacity not currently running but capable of being connected to the bus and load within a specified time.

nonutility generation. Electric generation by end users, or small power producers under PURPA, to supply electric power for industrial, commercial, and military operations, or sales to electric utilities.

nonutility power producer. A corporation, person, agency, authority, or other legal entity or instrumentality that owns or operates facilities for electric generation and is not an electric utility. Nonutility power producers include qualifying **cogenerators**, qualifying small power producers, and other nonutility generators (including IPPs). Nonutility power producers are without a designated franchised service area and do not file forms listed in the Code of Federal Regulations, Title 18, Part 141.

North American Electric Reliability Council (NERC). Electric utilities formed NERC to coordinate, promote, and communicate about the reliability of their generation and transmission systems. NERC is comprised of nine regional councils and one affiliate that together encompass most of the electric utility systems in the United States, Canada, and the northern portion of Baja California, Mexico. NERC reviews the overall reliability of existing and planned generation systems, sets reliability standards, and gathers data on demand, availability, and performance. The NERC regions are East Central Area Reliability Coordination Agreement (ECAR), Electric Reliability Council of Texas, Inc. (ERCOT), Florida Reliability Coordinating Council (FRCC), Mid-Atlantic Area Council (MAAC), Mid-America Interconnected Network, Inc. (MAIN), Midwest Reliability Organization (MRO), Northeast Power Coordinating Council (NPCC), Southeastern Electric Reliability Council (SERC), Southwest Power Pool, Inc. (SPP), and Western Electricity Coordinating Council (WECC).

North American Industry Classification System (NAICS). A new classification scheme, developed by the Office of Management and Budget to replace the Standard Industrial Classification (SIC) system, that categorizes establishments according to the types of production processes they primarily use.

nuclear fuel. Fissionable materials that have been enriched to such a composition that, when placed in a nuclear reactor, will support a self-sustaining fission chain reaction, producing heat in a controlled manner for process use.

nuclear power. Electricity generated by the use of the thermal energy released from the fission of nuclear fuel in a reactor.

nuclear reactor. An apparatus in which a nuclear fission chain reaction can be initiated, controlled, and sustained at a specific rate. A reactor includes fuel (fissionable material), moderating material to control the rate of fission, a heavy-walled pressure vessel to house reactor components, shielding to protect personnel, a system to conduct heat away from the reactor, and instrumentation for monitoring and controlling the reactor's systems.

Nuclear Regulatory Commission (NRC). Licenses operators of nuclear power plants. Reactor operators are authorized to control equipment that affects the power of the reactor in a nuclear power plant. In addition, an NRC-licensed senior reactor operator must be on duty during each shift to act as the plant supervisor and supervise the operations of all controls in the control room.

O

ocean energy. Draws on the energy of ocean waves, tides, or on the thermal energy (heat) stored in the ocean. Ocean energy technologies include wave energy, **tidal energy**, and **ocean thermal energy conversion systems**.

ocean energy systems. Energy conversion technologies that harness the energy in tides, waves, and thermal gradients in the oceans.

ocean thermal energy conversion (OTEC). The process or technologies for producing energy by harnessing the temperature differences (thermal gradients) between ocean surface waters and that of ocean depths. Warm surface water is pumped through an evaporator containing a working fluid in a closed Rankine-cycle system. The vaporized fluid drives a turbine/generator.

off-peak. Period of relatively low system demand. These periods often occur in daily, weekly, and seasonal patterns; these off-peak periods differ for each individual electric utility.

ohm. A measure of the electrical resistance of a material equal to the resistance of a circuit in which the potential difference of 1 V produces a current of 1 A.

Ohm's law. In a given electrical circuit, the amount of current in amperes is equal to the pressure in volts divided by the resistance in ohms. The principle is named after the German scientist Georg Simon Ohm.

on-peak. Periods of relatively high system demand. These periods often occur in daily, weekly, and seasonal patterns; these on-peak periods differ for each individual electric utility.

open access. A regulatory mandate to allow others to use a utility's transmission and distribution facilities to move bulk power from one point to another on a nondiscriminatory basis for a cost-based fee.

P

partial requirements consumer. A wholesale consumer with generating resources insufficient to carry all its load and whose energy seller is a long-term firm power source supplemental to the consumer's own generation or energy received from others. The terms and conditions of sale are similar to those for a full requirements consumer.

passive solar heating. A solar heating system that uses no external mechanical power, such as pumps or blowers, to move the collected solar heat.

peak demand. The maximum demand during a specified period of time.

peak load. The maximum load during a specified period of time.

peaking capacity. Capacity of generating equipment normally reserved for operation during the hours of highest daily, weekly, or seasonal loads. Some generating equipment may be operated at certain times as peaking capacity and at other times to serve loads on an around-the-clock basis.

permanently discharged fuel. Spent nuclear fuel for which there are no plans for reinsertion in the reactor core.

photovoltaic cell (PVC). An electronic device consisting of layers of semiconductor materials fabricated to form a junction (adjacent layers of materials with different electronic characteristics) and electrical contacts and being capable of converting incident light directly into electricity (DC).

photovoltaic module. An integrated assembly of interconnected photovoltaic cells designed to deliver a selected level of working voltage and current at its output terminals, packaged for protection against environmental degradation, and suited for incorporation in photovoltaic power systems.

plutonium (Pu). A heavy, fissionable, radioactive, metallic element (atomic number 94) that occurs naturally in trace amounts. It can also result as a by-product of the fission reaction in a uranium-fuel nuclear reactor and can be recovered for future use.

pole-mile. A unit of measuring the simple length of an electric transmission/distribution line/feeder carrying electric conductors, without regard to the number of conductors carried.

pole/tower type. The type of transmission line supporting structure.

power. An electric measurement unit of power called a volt-ampere is equal to the product of 1 V and 1 A. This is equivalent to 1 W for a DC system, and a unit of apparent power is separated into real and reactive power. Real power is the work-producing part of apparent power that measures the rate of supply of energy and is denoted as kilowatts (kW). Reactive power is the portion of apparent power that does no work and is referred to as kilovars; this type of power must be supplied to most types of magnetic equipment, such as motors, and is supplied by generator or by electrostatic equipment. Volt-amperes are usually divided by 1,000 and called kilovolt-amperes (kVA). Energy is denoted by the product of real power and the length of time utilized; this product is expressed as kilowatt-hours.

power distributors and dispatchers. Also called load dispatchers or systems operators, these workers control the flow of electricity through transmission lines to industrial plants and substations that supply residential electric needs. They monitor and operate current converters, voltage transformers, and circuit breakers. Dispatchers also monitor other distribution equipment and record readings at a pilot board—a map of the transmission grid system showing the status of transmission circuits and connections with substations and industrial plants. They also anticipate power needs, such as those caused by changes in the weather. They call control room operators to start or stop boilers and generators to bring production into balance with needs. Dispatchers handle emergencies, such as transformer or transmission line failures and route current around affected areas. In substations, they also operate and monitor equipment that increases or decreases voltage, and they operate switchboard levers to control the flow of electricity in and out of the substations.

power factor. The ratio of real power (kilowatt) to apparent power kilovolt-amperes for any given load and time.

power marketers. Business entities engaged in buying and selling electricity. Power marketers do not usually own generating or transmission facilities. Power marketers, as opposed to brokers, take ownership of the electricity and are involved in interstate trade. These entities file with FERC for status as a power marketer.

power plant operators. Control and monitor boilers, turbines, generators, and auxiliary equipment in power generating plants. Operators distribute power demands among generators, combine the current from several generators, and monitor instruments to maintain voltage and regulate electricity flows from the plant. When power requirements change, these workers start or stop generators and connect or disconnect them from circuits. They often use computers to keep records of switching operations and loads on generators, lines, and transformers. Operators may also use computers to prepare reports of unusual incidents, malfunctioning equipment, or maintenance performed during their shift.

power pool. An association of two or more interconnected electric systems having an agreement to coordinate operations and planning for improved reliability and efficiencies.

power production plant. All the land and land rights, structures and improvements, boiler or reactor vessel equipment, engines and engine-driven generator, turbogenerator units, accessory electric equipment, and miscellaneous power plant equipment are grouped together for each individual facility.

power system stabilizers (PSS). In recent years, a large amount of generation has been built in areas well suited for generation siting, but weakly connected to load centers or other parts of the transmission system. In such situations, small signal oscillations can be of concern. Properly tuned power system stabilizers may be required on machines in these generation-rich areas to minimize power oscillations between interconnected machines following a system fault. They become very important during real-time operation of the system since they can, and often do, determine the real-time export limit of the generation pocket. Additional importance must be placed on the plant and system operator awareness of whether the PSS is turned on during plant operation. A PSS that is not in service may lower the area export limit. If system operators are not aware that a PSS is off, the system can be inadvertently operated outside of the stability export limit. Under these conditions, a single system event in the area can lead to a widespread system outage.

powerhouse. A structure at a hydroelectric plant site that contains the turbine and generator.

pressurized water reactor (PWR). A nuclear reactor in which heat is transferred from the core to a heat exchanger via water kept under high pressure, so that high temperatures can be maintained in the primary system without boiling the water. Steam is generated in a secondary circuit.

primary energy. All energy consumed by end users, excluding electricity but including the energy consumed at electric utilities to generate electricity. (In estimating energy expenditures, there are no fuel-associated expenditures for hydroelectric power, geothermal energy, solar energy, or wind energy, and the quantifiable expenditures for process fuel and intermediate products are excluded.)

prime mover. The engine, turbine, water wheel, or similar machine that drives an electric generator; or, for reporting purposes, a device that converts energy to electricity directly (e.g., photovoltaic solar and fuel cells).

public utility. Providing essential public services, such as electric, gas, telephone, water, and sewer under legally established monopoly conditions.

public utility district (PUD). Municipal corporations organized to provide electric service to both incorporated cities and towns and unincorporated rural areas.

Public Utility Holding Company Act (PUHCA). Passed in 1935 as part of the New Deal legislation, and repealed on August 8, 2005 as part of a sweeping energy policy reform law, PUHCA's original intent was to halt corruption and scandals in the electric utility industry during the Great Depression era. It was meant to protect consumers against business dealings that often threatened the reliability of electric utilities. Specifically, it enabled extensive regulation of the size, spread, business type, and finances of holding companies owning and operating electric utility companies. Under the strictures of PUHCA, any companies seeking to become owners of public utilities had to divest themselves of their nonutility holdings. The rules were designed to make it extremely difficult for energy holding companies to get involved in risky businesses. With the passage of the 2005 Energy Policy Act, there are now no restrictions on who can buy public utilities. In addition, holding companies will not have to divest their nonutility businesses, and geographic limitations and restrictions on the number of holdings are released. Furthermore, even the SEC has been taken out of the review process. The SEC's role has been replaced with a much less strict review by FERC.

Public Utility Regulatory Policies Act of 1978 (PURPA). Among other things, PURPA promotes energy efficiency and increased use of alternative energy sources by encouraging companies to build cogeneration facilities and renewable energy projects using wind power, solar energy, geothermal energy, hydropower, biomass, and waste fuels. Facilities meeting PURPA's requirements are called QFs. PURPA encourages QF construction by requiring utilities to purchase QF electric output at a price no greater than the cost the utility would have incurred had it supplied the power itself or obtained it from another source. [Author's note: The Energy Policy Act of 2005, signed into law August 8, 2005, established conditions for eliminating PURPA's mandatory purchase obligation and revising the criteria for new QFs that seek to sell power under the mandatory purchase obligation. This obligation, according to some in the industry, costs consumers billions of dollars.]

pumped-storage hydroelectric plant. A plant that usually generates electric energy during peak load periods by using water previously pumped into an elevated storage reservoir during off-peak periods when excess generating capacity is available to do so. When additional generating capacity is needed, the water can be released from the reservoir through a conduit to turbine generators located in a power plant at a lower level.

purchased power. Power purchased or available for purchase from a source outside the system.

purchased power adjustment. A clause in a rate schedule that provides for adjustments to the bill when energy from another electric system is acquired and its cost varies from a specified unit base amount.

Q

qualifying facility (QF). A cogeneration or small power production facility that meets certain ownership, operating, and efficiency criteria established by FERC pursuant to PURPA.

quantity wires charge. A fee for moving electricity over the transmission and/or distribution system that is based on the quantity of electricity that is transmitted.

R

radioactive waste. Materials left over from making nuclear energy. Radioactive waste can destroy living organisms if it is not stored safely.

radioactivity. The spontaneous emission of radiation from the nucleus of an atom. Radionuclides lose particles and energy through this process.

railroad and railway electric service. Electricity supplied to railroads and interurban and street railways, for general railroad use, including the propulsion of cars or locomotives, where such electricity is supplied under separate and distinct rate schedules.

Rankine cycle. The thermodynamic cycle that is an ideal standard for comparing performance of heat engines, steam power plants, steam turbines, and heat pump systems that use a condensable vapor as the working fluid. Efficiency is measured as work done divided by sensible heat supplied.

Rankine cycle engine. The Rankine cycle system uses a liquid that evaporates when heated and expands to produce work, such as turning a turbine, which when connected to a generator, produces electricity. The exhaust vapor expelled from the turbine condenses, and the liquid is pumped back to the boiler to repeat the cycle. The working fluid most commonly used is water, though other liquids can also be used. Rankine cycle design is used by most commercial electric power plants. The traditional steam locomotive is also a common form of the Rankine cycle engine. The Rankine engine itself can be either a piston engine or a turbine.

rate base. The value of property upon which a utility is permitted to earn a specified rate of return as established by a regulatory authority. The rate base generally represents the value of property used by the utility in providing service and may be calculated by any one or a combination of the following accounting methods: fair value, prudent investment, reproduction cost, or original cost. Depending on which method is used, the rate base includes cash, working capital, materials and supplies, deductions for accumulated provisions for depreciation, contributions in aid of construction, customer advances for construction, accumulated deferred income taxes, and accumulated deferred investment tax credits.

rates. The authorized charges per unit or level of consumption for a specified time period for any of the classes of utility services provided to a customer.

rating. A manufacturer's guaranteed performance of a machine, transmission line, or other electrical apparatus, based on design features and test data. The rating will specify such limits as load, voltage, temperature, and frequency. The rating is generally printed on a nameplate attached to equipment and is commonly referred to as the nameplate rating or nameplate capacity.

reactive power. The electrical power that oscillates between the magnetic field of an inductor and the electrical field of a capacitor. Reactive power is never converted to nonelectrical power. It is calculated as the square root of the difference between the square of the kilovolt-amperes and the square of the kilowatts and is expressed as reactive volt-amperes.

refuse-derived fuel (RDF). A fuel produced by shredding municipal solid waste (MSW). Noncombustible materials such as glass and metals are generally removed prior to making RDF. The residual material is sold as-is or compressed into pellets, bricks, or logs. RDF processing facilities are typically located near a source of MSW, while the RDF combustion facility can be located elsewhere.

regional transmission group (RTG). A utility industry concept that FERC embraced for the certification of voluntary groups that would be responsible for transmission planning and use on a regional basis.

regulated entity. For the purpose of EIA's data collection efforts, entities that either provide electricity within a designated franchised service area and/or file forms listed in the Code of Federal Regulations, Title 18, part 141 are considered regulated entities. This includes electric IOUs that are subject to rate regulation, municipal utilities, federal and state power authorities, and rural electric cooperatives. Facilities that qualify as cogenerators or small power producers under PURPA are not considered regulated entities.

regulation, procedures, and practice. A utility commission carries out its regulatory functions through rulemaking and adjudication. Under rulemaking, the utility commission may propose a general rule of regulation change. By law, it must issue a notice of the proposed rule, and a request for comments is also made; the FERC publishes this in the Federal Register. The final decision must be published. A utility commission may also work on a case-by-case basis from submissions from regulated companies or others. Objections to a proposal may come from the commission or intervenors, in which case the proposal must be presented to a hearing presided over by an administrative law judge. The judge's decision may be adopted, modified, or reversed by the utility commissioners, in which case those involved can petition for a rehearing and may appeal a decision through the courts system to the U.S. Supreme Court.

reliability. There are two basic functional aspects of the term *reliability* as it relates to the interconnected bulk electric system. The first is adequacy and the second is security. Adequacy is the ability of the electric system to supply the aggregate electrical demand and energy requirements of customers at all times, taking into account scheduled and reasonably expected unscheduled outages of system elements. Security is the ability of the electric system to withstand sudden disturbances, such as electric short circuits or unanticipated loss of system elements.

reliability standards. Standards by which the object is to limit the number and duration of instances where load would exceed available capacity. Limits can be expressed in terms of the total number of hours per year on average that load would have to be curtailed because it exceeded available capacity. The optimal yearly duration for load shedding can be found by equating, at the margin, an annualized social cost of load shedding with the annual fixed cost (FC) of a generating plant (i.e., peaker unit) that reduces the quantity of load to be curtailed. The social cost of shedding load (in dollars per megawatt-hours) is the value of lost load (in dollars per megawatt-hours curtailed) multiplied by the duration of curtailment (in hours). Optimal duration of curtailment can be found by dividing the fixed cost of the generator by the value of lost load.

renewable energy resources. Energy resources that are naturally replenishing but flow-limited. They are virtually inexhaustible in duration but limited in the amount of energy that is available per unit of time. Renewable energy resources include: biomass, hydro, geothermal, solar, wind, ocean thermal, wave action, and tidal action.

repowering. Refurbishment of a plant by replacement of the combustion technology with a new combustion technology, usually resulting in better performance and greater capacity.

reregulation. The design and implementation of regulatory practices to be applied to the remaining regulated entities after restructuring of the vertically-integrated electric utility. The remaining regulated entities would be those that continue to exhibit characteristics of a natural monopoly, where imperfections in the market prevent the realization of more competitive results, and where, in light of other policy considerations, competitive results are unsatisfactory in one or more respects. Regulation could employ the same or different regulatory practices as those used before restructuring.

reserve generating capacity. Amount of generating capacity available to meet peak or abnormally high demands for power and to generate power during scheduled or unscheduled outages.

reserve margin (operating). The amount of unused available capability of an electric power system (at peak load for a utility system) as a percentage of total capability.

residential consumers. Consumers using gas for heating, air conditioning, cooking, water heating, and other residential uses in single and multifamily dwellings and apartments and mobile homes.

Residential Energy Consumption Survey (RECS). A national multistage probability sample survey conducted by the Energy End Use Division of the EIA. The RECS provides baseline information on how households in the United States use energy. The Residential Transportation Energy Consumption Survey (RTECS) sample is a subset of the RECS. Household demographic characteristics reported in the RTECS publication are collected during the RECS personal interview.

residential sector. An energy-consuming sector that consists of living quarters for private households. Common uses of energy associated with this sector include space heating, water heating, air conditioning, lighting, refrigeration, cooking, and running a variety of other appliances. The residential sector excludes institutional living quarters.

residential/commercial (consumer category). Housing units; wholesale or retail businesses (except coal wholesale dealers); health institutions (hospitals); social and educational institutions (schools and universities); and federal, state, and local governments (military installations, prisons, office buildings, etc.). Excludes shipments to federal power projects, such as TVA, and rural electrification cooperatives, power districts, and state power projects.

restructuring. The process of replacing a monopoly system of electric utilities with competing sellers, allowing individual retail customers to choose their electricity supplier but still receive delivery over the power lines of the local utility. It includes the reconfiguration of the vertically-integrated electric utility.

retail wheeling. The process of moving electric power from a point of generation across third-party-owned transmission and distribution systems to a retail customer.

reversible turbine. A hydraulic turbine, normally installed in a pumped-storage plant, which can be used alternatively as a pump or as an engine, turbine, water wheel, or other apparatus that drives an electrical generator.

Rural Electrification Administration (REA). A lending agency of the U. S. Department of Agriculture, the REA makes self-liquidating loans to qualified borrowers to finance electric and telephone service to rural areas. The REA finances the construction and operation of generating plants, electric transmission and distribution lines, or systems for the furnishing of initial and continued adequate electric services to persons in rural areas not receiving central station service.

S

sales for resale. A type of wholesale sales covering energy supplied to other electric utilities, cooperatives, municipalities, and federal and state electric agencies for resale to ultimate consumers.

sales to end users. Sales made directly to the consumer of the product. Includes bulk consumers, such as agriculture, industry, and utilities, as well as residential and commercial consumers.

scheduled outage. The shutdown of a generating unit, transmission line, or other facility for inspection or maintenance, in accordance with an advance schedule.

scheduling coordinators. Entities certified by FERC that act on behalf of generators, supply aggregators (wholesale marketers), retailers, and customers to schedule the distribution of electricity.

seasonal energy efficiency ratio (SEER). A measurement of efficiency for cooling devices such as heat pumps and air conditioners. A unit's SEER is calculated by dividing the total number of BTUs of heat removed from the air by the total amount of energy required by the unit. The higher the ratio, the more efficient the unit.

seasonal rates. Different seasons of the year are structured into an electric rate schedule whereby an electric utility provides service to consumers at different rates. The electric rate schedule usually takes into account demand based on weather and other factors.

securitization. A proposal for issuing bonds that would be used to buy down existing power contracts or other obligations. The bonds would be repaid by designating a portion of future customer bill payments. Customer bills would be lowered, since the cost of bond payments would be less than the power contract costs that would be avoided.

self-generator. A plant whose primary product is not electric power, but does generate electricity for its own use or for sale on the grid; for example, industrial combined heat and power plants.

semiconductor. Any material that has a limited capacity for conducting an electric current. Certain semiconductors, including silicon, gallium arsenide, copper indium diselenide, and cadmium telluride, are uniquely suited to the photovoltaic conversion process.

separate metering. Measurement of electricity or natural gas consumption in a building using a separate meter for each of several tenants or establishments in the building.

single-phase power. Homes use single-phase power, which is brought into houses by either aerial or buried through three wires. Two of these wires are covered with insulation and carry the power, while the third (often bare) is the ground wire. Before entering the house, the wires go through a watt-hour meter. On entering the home, the wires are fed to a circuit breaker or fuse box. This contains a disconnect switch to isolate the home from the power line, a main fuse, or circuit breaker, and breakers for the various circuits in the house. Separate breakers protect the 230-V lines for large appliances. Most homes are equipped with three wire connections to each outlet to provide full grounding protection.

small power producer (SPP). Under PURPA, a small power production facility (or small power producer) generates electricity using waste renewable (biomass, conventional hydroelectric, wind and solar, and geothermal) energy as a primary energy source. Fossil fuels can be used, but renewable resources must provide at least 75% of the total energy input. (See Code of Federal Regulations, Title 18, Part 292.)

solar energy. The radiant energy of the Sun, which can be converted into other forms of energy, such as heat or electricity.

spent fuel. Irradiated fuel that is permanently discharged from a reactor. Except for possible reprocessing, this fuel must eventually be removed from its temporary storage location at the reactor site and placed in a permanent repository. Spent fuel is typically measured either in metric tons of heavy metal (i.e., only the heavy metal content of the spent fuel is considered) or in metric tons of initial heavy metal (essentially, the initial mass of the fuel before irradiation). The difference between these two quantities is the weight of the fission products.

spinning reserve. That reserve generating capacity running at a zero load and synchronized to the electric system.

stand-alone generator. A power source/generator that operates independently of (or is not connected to) an electric transmission and distribution network; used to meet a load or loads physically close to the generator.

standby electricity generation. Involves use of generators during times of high demand on utilities to avoid extra peak-demand charges.

standby facility. A facility that supports a utility system and is generally running under no load. It is available to replace or supplement a facility normally in service.

static reactive power devices. Used to maintain steady-state voltage levels. An example is switched capacitor banks. Static reactive power compensation can be thought of as keeping reactive power available in the generators by supplying the reactive power consumption needs of heavily loaded transmission lines. Increasing the amount of reactive power available in an area can also be accomplished by building new transmission lines to relieve the loading on the existing transmission grid. This will have the effect of reducing the reactive power consumption by the transmission system, which will help to support system voltage levels and real power transfer to loads. Compare with **dynamic reactive power sources**.

steam boiler. Type of furnace in which fuel is burned and the heat is used to produce steam.

steam power plants. A fossil-fuel steam plant operation consists of four essential steps. First, water is pumped at high pressure to a boiler. Next it is heated by fossil-fuel combustion to produce steam at high temperature and pressure. In the third step, this steam flows through a turbine, rotating an electric generator (connected to the turbine shaft), which converts the mechanical energy to electricity. The final step is that the turbine exhaust steam is condensed by using cooling water from an external source to remove the heat rejected in the condensing process. The condensed water is pumped back to the boiler to repeat the cycle.

steam turbine. A device that converts high-pressure steam, produced in a boiler, into mechanical energy that can then be used to produce electricity by forcing blades in a cylinder to rotate and turn a generator shaft.

stranded benefits. Benefits associated with regulated retail electric service that may be at risk under open market retail competition. Examples include conservation programs, fuel diversity, reliability of supply, and tax revenues based on utility revenues.

stranded costs. Costs incurred by a utility that may not be recoverable under market-based retail competition. Examples include undepreciated generating facilities, deferred costs, and long-term contract costs.

subbituminous coal. A coal whose properties range from those of lignite to those of bituminous coal and used primarily as fuel for steam-electric power generation. It may be dull, dark brown to black, and soft and crumbly at the lower end of the range, and bright, jet black, hard, and relatively strong at the upper end. Subbituminous coal contains 20% to 30% inherent moisture by weight. The heat content of subbituminous coal ranges from 17 to 24 million BTU per ton on a moist, mineral-matter-free basis. The heat content of subbituminous coal consumed in the United States averages 17 to 18 million BTU per ton, on the as-received basis (i.e., containing both inherent moisture and mineral matter).

substation. Facility equipment that switches, changes, or regulates electric voltage.

subtransmission. A set of transmission lines of voltages between transmission voltages and distribution voltages. Generally, these lines are in the voltage range of 69 kV to 138 kV.

sulfur dioxide (SO_2). A toxic, irritating, colorless gas soluble in water, alcohol, and ether. Used as a chemical intermediate, in paper pulping and ore refining, and as a solvent.

surplus energy. Energy generated that is beyond the immediate needs of the producing system. This energy may be supplied by spinning reserves and sold on an interruptible basis.

switching station. Facility equipment used to tie together two or more electric circuits through switches. The switches are selectively arranged to permit a circuit to be disconnected or to change the electric connection between the circuits.

system interconnection. A physical connection between two electric systems that permits the transfer of electric energy in either direction.

T

Tennessee Valley Authority (TVA). A federal agency established in 1933 to develop the Tennessee river Valley region of the southeastern United States.

terawatt-hour. This is 1 trillion watt-hours.

thermal. A term used to identify a type of electric generating station, capacity, capability, or output in which the source of energy for the prime mover is heat.

thermal efficiency. A measure of the efficiency of converting a fuel to energy and useful work; useful work and energy output divided by higher heating value of input fuel times 100 (for percent).

thermal energy storage. The storage of heat energy during utility off-peak times at night, for use during the next day without incurring daytime peak electric rates.

thermal limit. The maximum amount of power a transmission line can carry without suffering heat-related deterioration of line equipment, particularly conductors.

thermal resistance (R-value). This designates the resistance of a material to heat conduction. The greater the R-value, the larger the number.

thermal storage. Storage of heat or heat sinks (coldness) for later heating or cooling. Examples are the storage of solar energy for night heating, the storage of summer heat for winter use, the storage of winter ice for space cooling in the summer, and the storage of electrically-generated heat or coolness when electricity is less expensive, to be released in order to avoid using electricity when the rates are higher. There are four basic types of thermal storage systems: ice storage; water storage; storage in rock, soil or other types of solid thermal mass; and storage in other materials, such as glycol (antifreeze).

thermophotovoltaic cell. A device where sunlight concentrated onto an absorber heats it to a high temperature. The thermal radiation emitted by the absorber is used as the energy source for a photovoltaic cell that is designed to maximize conversion efficiency at the wavelength of the thermal radiation.

third-party DSM program sponsor. An energy service company (ESCO) that promotes a program sponsored by a manufacturer or distributor of energy products such as lighting or refrigeration whose goal is to encourage consumers to improve energy efficiency, reduce energy costs, change the time of usage, or promote the use of a different energy source.

three-phase power. Power generated and transmitted from generator to load on three conductors.

tidal energy. An ocean energy technology. Traditionally involves erecting a dam across the opening to a tidal basin. The dam includes a sluice that is opened to allow the tide to flow into the basin; the sluice is then closed, and as the sea level drops, traditional hydropower technologies can be used to generate electricity from the elevated water in the basin. Some researchers are trying to extract energy directly from tidal flow streams.

tie-line. A transmission line connecting two or more power systems.

topping cycle. A boiler produces steam to power a turbine-generator to produce electricity. The steam leaving the turbine is used in thermal applications such as space heating and/or cooling or is delivered to other end user(s).

transfer capability. The overall capacity of interregional or international power lines, together with the associated electrical system facilities, to transfer power and energy from one electrical system to another.

transformer. An electrical device for changing the voltage of AC.

transmission. The movement or transfer of electric energy over an interconnected group of lines and associated equipment between points of supply and points at which it is transformed for delivery to consumers or is delivered to other electric systems. Transmission is considered to end when the energy is transformed for distribution to the consumer.

transmission and distribution loss. Electric energy lost due to the transmission and distribution of electricity. Much of the loss is thermal in nature.

transmission circuit. A conductor used to transport electricity from generating stations to load.

transmission line. A set of conductors, insulators, supporting structures, and associated equipment used to move large quantities of power at high voltage, usually over long distances between a generating or receiving point and major substations or delivery points.

transmission network. A system of transmission or distribution lines so cross-connected and operated as to permit multiple power supply to any principal point.

transmission system. An interconnected group of electric transmission lines and associated equipment for moving or transferring electric energy in bulk between points of supply and points at which it is transformed for delivery over the distribution system lines to consumers or is delivered to other electric systems.

transmitting utility. A regulated entity that owns and may construct and maintain wires used to transmit wholesale power. It may or may not handle the power dispatch and coordination functions. It is regulated to provide nondiscriminatory connections, comparable service, and cost recovery. According to the EPAct of 1992, it includes any electric utility, qualifying cogeneration facility, qualifying small power production facility, or federal power marketing agency that owns or operates electric power transmission facilities that are used for the sale of electric energy at wholesale.

turbine. A machine for generating rotary mechanical power from the energy of a stream of fluid (such as water, steam, or hot gas). Turbines convert the kinetic energy of fluids to mechanical energy through the principles of impulse and reaction, or a mixture of the two.

U

ultimate customer. A customer that purchases electricity for its own use and not for resale.

ultraviolet. Electromagnetic radiation in the wavelength range of 4 to 400 nanometers.

unbundling. Separating vertically integrated monopoly functions into their component parts for the purpose of separate service offerings.

unregulated entity. For the purpose of EIA's data collection efforts, entities that do not have a designated franchised service area and that do not file forms listed in the Code of Federal Regulations, Title 18, Part 141, are considered unregulated entities. This includes qualifying cogenerators, qualifying small power producers, and other generators that are not subject to rate regulation, such as IPPs.

unscheduled outage service. Power received by a system from another system to replace power from a generating unit forced out of service.

uranium (U). A heavy, naturally radioactive, metallic element (atomic number 92). Its two principally occurring isotopes are uranium-235 and uranium-238. Uranium-235 is indispensable to the nuclear industry because it is the only isotope existing in nature, to any appreciable extent, that is fissionable by thermal neutrons. Uranium-238 is also important because it absorbs neutrons to produce a radioactive isotope that subsequently decays to the isotope plutonium-239, which also is fissionable by thermal neutrons.

useful thermal output. The thermal energy made available in a CHP system for use in any industrial or commercial process, heating or cooling application, or delivered to other end users, i.e., total thermal energy made available for processes and applications other than electrical generation.

utility distribution companies. The entities that will continue to provide regulated services for the distribution of electricity to customers and serve customers who do not choose direct access. Regardless of where a consumer chooses to purchase power, the customer's current utility, also known as the utility distribution company, will deliver the power to the consumer.

utility DSM costs. The costs incurred by the utility to achieve the capacity and energy savings from the DSM program. Costs incurred by consumers or third parties are to be excluded. The costs are to be reported in nominal dollars in the year in which they are incurred, regardless of when the savings occur. The utility costs are all the annual expenses (labor, administrative, equipment, incentives, marketing, monitoring and evaluation, and other) incurred by the utility for operation of the DSM program, regardless of whether the costs are expensed or capitalized. Lump-sum capital costs (typically accrued over several years prior to start up) are not to be reported. Program costs associated with strategic load growth activities are also to be excluded.

utility generation. Generation by electric systems engaged in selling electric energy to the public.

V

variable-speed wind turbines. Turbines in which the rotor speed increases and decreases with changing wind speed, producing electricity with a variable frequency.

vertical integration. The combination within a firm or business enterprise of one or more stages of production or distribution. In the electric industry, it refers to the historical arrangement whereby a utility owns its own generating plants, transmission system, and distribution lines to provide all aspects of electric service.

vertical-axis wind turbine (VAWT). A type of wind turbine in which the axis of rotation is perpendicular to the wind stream and the ground.

volt (V). The volt is the International System of Units (SI) measure of electric potential or electromotive force. A potential of 1 V appears across a resistance of 1 ohm when a current of 1 A flows through that resistance. Reduced to SI base units, $1 \text{ V} = 1 \text{ kg} \times \text{m}^2 \times \text{s}^{-3} \times \text{A}^{-1}$ (kilogram-meter squared per second cubed per ampere).

voltage. The difference in electrical potential between any two conductors or between a conductor and ground. It is a measure of the electric energy per electron that electrons can acquire and/or give up as they move between the two conductors.

voltage reduction. Any intentional reduction of system voltage by 3% or greater for reasons of maintaining the continuity of service of the bulk electric power supply system.

voltage regulation. Long transmission lines have considerable inductance and capacitance as well as resistance. When a current flows through the line, inductance and capacitance have the effect of varying the voltage on the line as the current varies. Thus the supply voltage varies with the load. Several kinds of devices are used to overcome this undesirable variation, in an operation called regulation of the voltage. They include induction regulators and three-phase synchronous motors (called synchronous condensers), both of which vary the effective amount of inductance and capacitance in the transmission circuit. Inductance and capacitance react with a tendency to nullify one another. When a load circuit has more inductive than capacitive reactance, as almost always occurs in large power systems, the amount of power delivered for a given voltage and current is less than when the two are equal. The ratio of these two amounts of power is called the *power factor*. Because transmission line losses are proportional to current, capacitance is added to the circuit when possible, thus bringing the power factor as nearly as possible to 1. For this reason, large capacitors are frequently inserted as part of power-transmission systems.

volt-ampere reactive (VAR). A reactive load, typically inductive from electric motors that causes more current to flow in the distribution network than is actually consumed by the load. This requires excess capability on the generation side and causes greater power losses in the distribution network.

W–Z

waste energy. Municipal solid waste, landfill gas, methane, digester gas, liquid acetonitrile waste, tall oil, waste alcohol, medical waste, paper pellets, sludge waste, solid by-products, tires, agricultural by-products, closed loop biomass, fish oil, and straw used as fuel.

waste heat boiler. A boiler that receives all or a substantial portion of its energy input from the combustible exhaust gases from a separate fuel-burning process.

waste heat recovery. Any conservation system whereby some space heating or water heating is done by actively capturing by-product heat that would otherwise be ejected into the environment. In commercial buildings, sources of water-heat recovery include refrigeration/air-conditioner compressors, manufacturing or other processes, data processing centers, lighting fixtures, ventilation exhaust air, and the occupants themselves. Not to be considered is the passive use of radiant heat from lighting, workers, motors, ovens, etc., when there are no special systems for collecting and redistributing heat.

waste materials. Otherwise discarded combustible materials that, when burned, produce energy for such purposes as space heating and electric power generation. The size of the waste may be reduced by shredders, grinders, or hammer mills. Noncombustible materials, if any, may be removed. The waste may be dried and then burned, either alone or in combination with fossil fuels.

water pollution abatement equipment. Equipment used to reduce or eliminate waterborne pollutants, including chlorine, phosphates, acids, bases, hydrocarbons, sewage, and other pollutants. Examples of water pollution abatement structures and equipment include those used to treat thermal pollution; cooling, boiler, and cooling tower blowdown water; coal pile runoff; and fly ash wastewater. Water pollution abatement excludes expenditures for treatment of water prior to use at the plant.

water turbine. A turbine that uses water pressure to rotate its blades; the primary types are the Pelton wheel, for high heads (pressure); the Francis turbine, for low to medium heads; and the Kaplan, for a wide range of heads. Primarily used to power an electric generator.

watt (W). The unit of electrical power equal to 1. A under a pressure of 1 V; 1 W is equal to 1/746 horsepower.

watt meter. A device for measuring power consumption.

watt-hour (Wh). The electrical energy unit of measure equal to 1 W of power supplied to, or taken from, an electric circuit steadily for 1 hour.

wave technology. An ocean energy technology. The total power of waves breaking on the world's coastlines as estimated by the DOE is 2 to 3 million MW. In favorable locations, wave energy density can average 65 MW per mile of coastline.

wheeling charge. An amount charged by one electrical system to transmit the energy of, and for, another system or systems.

wheeling service. The movement of electricity from one system to another over transmission facilities of interconnecting systems. Wheeling service contracts can be established between two or more systems.

wholesale competition. A system whereby a distributor of power would have the option to buy its power from a variety of power producers, and the power producers would be able to compete to sell their power to a variety of distribution companies.

wholesale electric power market. The purchase and sale of electricity from generators to resellers (retailers), along with the ancillary services needed to maintain reliability and power quality at the transmission level.

wholesale sales. Energy supplied to other electric utilities, cooperatives, municipals, and federal and state electric agencies for resale to ultimate consumers.

wholesale transmission services. The transmission of electric energy sold, or to be sold, in the wholesale electric power market.

wholesale wheeling. An arrangement in which electricity is transmitted from a generator to a utility through the transmission facilities of an intervening system.

wind energy. Uses the energy in the wind for practical purposes like generating electricity, charging batteries, pumping water, or grinding grain. Large, modern wind turbines operate together in wind farms to produce electricity for utilities. Small turbines are used by homeowners and remote villages to help meet energy needs. Wind energy is considered a *green power* technology.

wind energy conversion system. An apparatus for converting the energy available in the wind to mechanical energy that can be used to power machinery (grain mills, water pumps) and to operate an electrical generator.

wind energy technologies. Modern wind turbines are divided into major categories: horizontal axis turbines and vertical axis turbines. Old-fashioned windmills are still seen in many rural areas.

wind farm. A group of wind turbines interconnected to a common utility system through a system of transformers, distribution lines, and (usually) one substation. Operation, control, and maintenance functions are often centralized through a network of computerized monitoring systems, supplemented by visual inspection. This is a term commonly used in the United States. In Europe, it is called a generating station.

wind turbines. These turbines capture the wind's energy with two or three propeller-like blades, which are mounted on a rotor, to generate electricity. The turbines sit high atop towers, taking advantage of the stronger and less turbulent wind at 100 ft (30 m) or more above ground. A blade acts much like an airplane wing. When the wind blows, a pocket of low-pressure air forms on the downwind side of the blade. The low-pressure air pocket then pulls the blade toward it, causing the rotor to turn. This is called *lift*. The force of the lift is actually much stronger than the wind's force against the front side of the blade, which is called *drag*. The combination of lift and drag causes the rotor to spin, like a propeller, and the turning shaft spins a generator to make electricity. Wind turbines can be used as stand-alone applications, or they can be connected to a utility power grid or even combined with a photovoltaic (solar cell) system. Stand-alone turbines are typically used for water pumping or communications. However, homeowners and farmers in windy areas may also use turbines to generate electricity. For utility-scale sources of wind energy, a large number of turbines are usually built close together to form a wind farm.

wires charge. A broad term referring to fees levied on power suppliers or their customers for the use of the transmission or distribution wires.

Appendix

Acronyms and Abbreviations

A

A	ampere
AC	alternating current
ACA/CIPCA	American Cogeneration Association/Cogeneration and Independent Power Coalition of America
ACBM	asbestos-containing building material
AEE	Association of Energy Engineers
AFC	automatic frequency control
AFUDC	allowance for funds used during construction
AGA	American Gas Association
AIMS	Automated Interchange Matching System, Inc.
AM	automated mapping
AM/FM	automated mapping/facilities management
AMR	automatic meter reading
ANEC	American Nuclear Energy Council
ANS	American Nuclear Society
APA	Alaska Power Administration
APGA	American Public Gas Association
API	American Petroleum Institute
APM	affiliated power marketer
APPA	American Public Power Association
APU	auxiliary power unit
ASCC	Alaska Systems Coordinating Council
ASE	Alliance to Save Energy

ASES	American Solar Energy Society
ASTM	American Society for the Testing of Materials
ATC	available transmission capability
ATWT	atomic weight
AWEA	American Wind Energy Association

B

BACT	best available control technology
bbl	barrel
bcf	billion cubic feet
BERA	Biomass Energy Research Association
BLM	Bureau of Land Management
BLS	Bureau of Labor Statistics
BOE	barrels of oil equivalent
bp	boiling point
BPA	Bonneville Power Administration
BPL	broadband over power lines
BRPU	biennial resource planning update
BTU	British thermal unit
BTX	benzene, toluene, xylene
BWR	boiling water reactor

C

C	Celsius
CAA	Clean Air Act
CAAA	Clean Air Act Amendments of 1990
CAMR	Clean Air Mercury Rule
CARB	California Air Resources Board
CARE	Coalition for Affordable and Reliable Energy
CC plant	combined-cycle plant
CCCC	Climate Change Credit Corporation
CCPI	Clean Coal Power Initiative
CDD	cooling degree-days
CEC	California Energy Council
CECA	Consumer Energy Council of America
CEMS	Continuous Emission Monitoring System
CEQ	Council on Environmental Quality

CERA	Cambridge Energy Research Associates
CERTS	Consortium for Electricity Reliability Technology Solutions
cf	cubic foot
CFB	circulating fluidized bed
CFC	chlorofluorocarbon
CFC	Cooperative Finance Corporation
cfs	cubic feet per second
CFTC	Commodities Futures Trading Commission
CH_4	methane
CHP	combined heat and power (plant)
CLECA	California Large Energy Consumers Association
CNG	compressed natural gas
CO	carbon monoxide
CO_2	carbon dioxide
COE	Corps of Engineers (Army)
COPAS	Council of Petroleum Accountant Societies
CPEX	Continental Power Exchange Inc.
CPI	Consumer Price Index
CPO	commodity pool operator
CRT	capacity reservation tariff
CT	combustion turbine
CTL	coal-to-liquids
CWIP	construction work in progress

D

DA	distribution automation
DA/DSM	distribution automation/demand-side management
DC	direct current
DE	distributed energy
DOA	Department of Agriculture
DOE	Department of Energy
DOI	Department of the Interior
DOJ	Department of Justice
DPC	Domestic Petroleum Council
DPS	dividends per share
DSM	demand-side management

E

EBB	electronic bulletin board
e-bill	electronic bill
ECAR	East Central Area Reliability Coordination Agreement
ECRC	Electricity Consumers Resource Council
EEA	Energy and Environmental Analysis, Inc.
EEI	Edison Electric Institute
EGA/PCA	Electric Generation Association/Power Coalition of America
EGSA	Electrical Generation Systems Association
EHV	extra-high voltage
EIA	Energy Information Administration
EIN	Electronic Information Network
EIS	Environmental Impact Statement
EISB	Electric Industry Standards Board
EMCS	energy management and control systems
EMF	electric and magnetic fields
EMS	emergency management system
EPA	Environmental Protection Agency
EPC	engineering, procurement, and construction
EPGA	Electric Power Generation Association
EPRI	Electric Power Research Institute
EPS	earnings per share
EPSA	Electric Power Supply Association
ERA	Economic Regulatory Administration
ERAM	electric revenue adjustment mechanism
ERCOT	Electric Reliability Council of Texas, Inc.
ERDA	Energy Research and Development Administration
ERP	emergency restoration plan
ESCO	energy service company
ESP	energy service provider
ESPC	energy savings performance contracts
EWG	exempt wholesale generator

F

F	Fahrenheit
FACTS	flexible alternating current transmission system
FASB	Financial Accounting Standards Board

FBC	fluidized bed combustion
FBR	fast breeder reactor
FEC	Federal Energy Commission
FEIS	federal environmental impact statement
FERC	Federal Energy Regulatory Commission
FGD	flue gas desulfurization
FM	facilities management
FPA	Federal Power Act
FPC	Federal Power Commission
FRCC	Florida Reliability Coordinating Council
FV	fixed variable rate structure
FWS	Fish and Wildlife Service

G

G&T	generation and transmission cooperative
GAMA	Gas Appliance Manufacturers Association
GDP	gross domestic product
GIS	geographic information system
GISB	Gas Industry Standards Board
GITA	Geospatial Information and Technology Association
GPA	Gas Processors Association
GPS	global positioning system
GRC	general rate case
GRI	Gas Research Institute
GTL	gas-to-liquids
GW	gigawatt
GWe	gigawatt electric
GWh	gigawatt hour
GWP	global warming potential

H

HAPs	hazardous air pollutants
HDD	heating degree-days
HFCs	hydrofluorocarbons
hp	horsepower
HRSG	heat recovery steam generator
HTGR	high-temperature gas-cooled reactor

HVAC	heating, ventilating, and air conditioning
HVAC	high-voltage alternating current
HVDC	high-voltage direct current
Hz	hertz

I

IADC	International Association of Drilling Contractors
IB	introducing brokers
IC	internal combustion
IEA	International Energy Agency
IEEE	Institute of Electrical and Electronic Engineers
IGCC	integrated gasification combined cycle
INGAA	Interstate Natural Gas Association of America
INPO	Institute of Nuclear Power Operations
IOGCC	Interstate Oil and Gas Compact Commission
IOU	investor-owned utility
IP	Internet protocol (networks)
IPAA	Independent Petroleum Association of America
IPM	independent power marketer
IPP	independent power producer
IROR	incentive rate of return
IRP	integrated resource planning
IRS	Internal Revenue Service
ISO	independent system operator
IT	information technology
ITC	independent transmission companies

J

J	joule
JCM	joint and common market (electricity)
J/K	joules per Kelvin
J/s	joules per second
JCL	job control language
JFET	junction field-effect transistor
JMOS	junction metal oxide semiconductor

K

K	Kelvin
kHz	kilohertz
km	kilometer
kV	kilovolt
kVA	kilovolt-ampere
kW	kilowatt
kWe	kilowatt electric
kWh	kilowatt-hour

L

LAER	lowest achievable emission rate
LAN	local area network
LDC	local distribution company
LDs	liquidated damages
LFG	landfill gas
LIHEAP	Low-Income Home Energy Assistance Program
LNG	liquefied natural gas
LPG	liquefied petroleum gas
LPPC	large public power council
LRAC	long-run avoided cost
LWR	light water reactor (nuclear)

M

m	meter
M&A	mergers and acquisitions
MAAC	Mid-Atlantic Area Council
MACT	maximum achievable control technology
MAIN	Mid-America Interconnected Network
MAPP	Mid-Continent Area Power Pool
MBtu	Million British thermal units
MCA	Midwest Cogeneration Association
Mcf	thousand cubic feet
MEA	Midwest Energy Association
MFV	modified fixed-variable rates
MHz	megahertz

MISO	Midwest Independent Transmission System Operator, Inc.
MMbbl/d	one million barrels of oil per day
MMBtu	one million British thermal units
MMcf	one million cubic feet
MMgal/d	one million gallons per day
mpg	miles per gallon
MRO	Midwest Reliability Organization
MMS	Minerals Management Service
MMst	one million short tons
MOU	memorandum of understanding
mph	miles per hour
MPS	mobile power system
m/s	meters per second
MSW	municipal solid waste
MTBE	methyl tertiary butyl ether
MVA	megavolt-amperes
MW	megawatt
MWe	megawatt electric
MWh	megawatt-hour
MWt	megawatt thermal

N

NAAQS	1997 National Ambient Air Quality Standards
NAESCO	National Association of Energy Service Companies
NAICS	North American Industry Classification System
NARUC	National Association of Regulatory Utility Commissioners
NASEO	National Association of State Energy Officials
NDA	National Drilling Association
NEA	National Energy Act
NEI	Nuclear Energy Institute
NELA	National Electric Light Association
NEMA	National Electrical Manufacturers Association
NEMA	National Energy Marketers Association
NEPA	National Environmental Policy Act
NEPDG	National Energy Policy Development Group
NEPOOL	New England Power Pool
NERC	North American Reliability Council
NESHAP	National Emission Standards for Hazardous Air Pollutants

NGA	Natural Gas Act
NGL	natural gas liquids
NGNP	next-generation nuclear plant
NGPA	Natural Gas Policy Act of 1978
NGSA	Natural Gas Supply Association
NHA	National Hydropower Association
NISC	National Information Solutions Cooperative
NMA	National Mining Association
N_2O	nitrous oxide
NO_x	nitrogen oxides
NOAA	National Oceanic and Atmospheric Administration
NOI	Notice of Inquiry
NOPR	Notice of Proposed Rulemaking
NPC	National Petroleum Council
NPCC	Northeast Power Coordinating Council
NRC	Nuclear Regulatory Commission
NRDC	Natural Resources Defense Council
NRECA	National Rural Electric Cooperative Association
NRRI	National Regulatory Research Institute
NRTC	National Rural Telecommunications Cooperative
NSPS	new source performance standards
NSR	new source review
NSWA	National Stripper Well Association
NUG	nonutility generator
NURE	National Uranium Resource Evaluation
NWPA	Nuclear Waste Policy Act
NWTC	National Wind Technology Center
NYMEX	New York Mercantile Exchange

O

O&M	operations and maintenance
OASIS	open access same-time information systems
OCR	optical character reader
OCS	outer continental shelf
ODS	ozone depleting substance
OEM	original equipment manufacturer
OETD	Office of Electric Transmission and Distribution
OGS	off-gas system

OPEC	Organization of the Petroleum Exporting Countries
OPS	offshore power systems
OPS	operational protection system
OSC	operational support center
OSHA	Occupational Safety and Health Administration
OSS	operational storage site
OT	operating temperature
OTAG	ozone transport assessment group
OTEC	ocean thermal energy conversion
OVP	over-voltage protection

P

PBMR	pebble bed modular reactor
PBR	performance-based rates
PCAIR	Proposed Clean Air Interstate Rule
PEM	polymer electrolyte membrane
PFCs	perfluorocarbons
PGA	purchased gas adjustment
PJM	Pennsylvania/New Jersey/Maryland Interconnection
PLC	power-line carrier
PM	particulate matter
PMA	Power Marketing Association
PPA	purchase power agreement
ppb	parts per billion
PPE	power, plant and equipment (expenditures)
ppm	parts per million
PSA	power sales agreement
PSC	Public Service Commission
PSS	power system stabilizers
PTC	Production Tax Credit (renewable energy)
Pu	plutonium
PUC	public utility commission
PUD	public utility district
PUHCA	Public Utility Holding Company Act
PURPA	Public Utility Regulatory Policies Act
PV	photovoltaic (solar)
PVC	photovoltaic cell
PWR	pressurized water reactor (nuclear)

Q

Q	quality factor (of a resonant circuit)
QA	quality assurance
QAP	quality assurance program
QC	quality control
QF	qualifying facility
QOS	quality of service
QTP	quality test plan
quad	quadrillion BTU

R

R&D	research and development
RAC	refiners' acquisition costs
RACT	reasonably available control technology
RCRA	Resource Conservation and Recovery Act
RDF	refuse-derived fuel
REA	Rural Electrification Administration (see RUS)
RECS	residential energy consumption survey
RES	renewable electricity standards
RIE	recognized investment exchange
RIM	ratepayer impact measure test
RIN	real-time information network
ROC	regulatory-out clause
ROE	return on equity
ROW	right-of-way
R/P Ratio	reserves-to-production ratio
rpm	revolutions per minute
RPS	renewable portfolio standard
RTG	regional transmission group
RTO	regional transmission organization
RUS	Rural Utilities Service (as of May 2005, REA is known as RUS)

S

SCADA	supervisory control and data acquisition
SCF	standard cubic feet
SCR	selective catalytic reduction

SEC	Securities and Exchange Commission
SEER	Seasonal Energy Efficiency Ratio
SEIA	Solar Energy Industries Association
SEPA	Southeastern Power Administration
SERC	Southeastern Electric Reliability Council
SFV	straight fixed variable
SIB	Securities Investment Board
SIC	standard industrial classification
SIP	state implementation plan
SMP	special marketing plan
SMR	steam methane reformation
SNF	spent nuclear fuel
SNG	synthetic natural gas
SO$_2$	sulfur dioxide
SPP	Southwest Power Pool, Inc.
SPP	small power producer
SPS	special protection system
SRO	self-regulatory organization
SSA	steam sales agreement
SWPA	Southwestern Power Administration

T

t	ton
T	transmittance
T&D	transmission and distribution
TAVS	turbine-area ventilation system
TB	turbine building
TBCCW	turbine building closed cooling water
TBESI	turbine building exhaust system isolation
TBS	turbine by-pass system
TC	thermocouple
TC	temperature coefficient
TC	temperature controller
tcf	trillion cubic feet
TCG	total combustible gases
TCP	transmission control protocol
TCS	turbine control system
TCV	turbine control valve

t/d	tons per day
TDAC	time-differentiated rates with average costs
TDC	thermal diffusion coefficient
TDI	turbine disk integrity
TDMC	time-differentiated rates with marginal costs
TDS	total dissolved solids
TDS	time-delay switch
TE	transverse electric
T/E	thermoelectric
TEC	thermal electric coding
TEC	total estimated cost
TEG	thermoelectric generator
TEM	transmission electron microscope
TEM	transverse electromagnetic mode
TENR	technically enhanced naturally radioactive
TENRAP	technically enhanced naturally radioactive products
TES	thermal energy storage
TFC	total final consumption of energy
TFE	tetrafluoride ethylene
TFE	thermionic fuel element
TFS	turbine first stage
TFT	thin-film transistor
TG	turbine-generator
TGB	turbine-generator building
TGS	turbine-generator system
TGV	turbine governor valve
t/h	tons per hour
THD	total harmonic distortion
THz	terahertz
TI	temperature indicator
TIC	temperature indicating controller
TLO	turbine lube oil
TOD	time of day
TOP	transient over power
TOU	time-of-use tariffs
TPES	total primary energy supply
TPF	total peaking factor
TPM	thermal power monitor

TPV	thermophotovoltaic
TQC	total quality control
TQM	total quality management
TPY	tons per year
TRC	total resource cost test
TRI	toxic release inventory
tsi	tons per square inch
TSIN	Transmission Services Information Network
TSP	total suspended particulates
TSV	turbine stop valves
TSW	turbine-building service water
TTC	total transmission capability
TTC	time-to-collection (data acquisition)
TTE	thermal transient equipment
tU	tons of uranium
TVA	Tennessee Valley Authority
TVR	Tennessee Valley region
TW	terawatt
TWh	terawatt-hours

U

U	uranium
UA	utility automation
UART	universal asynchronous receiver-transmitter
UCS	Union of Concerned Scientists
UDS	underground distribution system
UHF	ultra-high frequency
UHV	ultra-high voltage
ULD	upper load demand
ULEV	ultra low emission vehicle
UNI	user network interface
UOP	unit operating procedure
UPH	underground pumped hydro
UPS	uninterruptible power supply
UR	user requirements
URD	underground residential distribution
USEA	United States Energy Association
UTC	Utilities Telecommunications Commission

V

V	volt
VA	volt-ampere
VAC	volts alternating current
VAR	volt-ampere reactive
VAR	value-added reseller
VAWT	vertical-axis wind turbine
VCO	voltage-controlled oscillator
VCXO	voltage-controlled crystal oscillator
VDC	volts direct current
VFC	voltage-to-frequency converter
VHF	very-high frequency
VHSIC	very high speed integrated circuit
VHTGR	very high temperature gas reactor
VHTR	very high temperature gas-cooled reactor
VLF	very-low frequency
VOC	volatile organic compounds
VRA	vulnerability assessment plan
VSWR	voltage standing-wave ratio
VTCO	voltage-temperature cutoff

W

W	watt
WAIS	wide area information server
WAMS	wide area monitoring systems
WAN	wide area network
WAPA	Western Area Power Administration
WDC	waste disposal cask
WDS	waste disposal system
WECC	Western Electricity Coordinating Council
WECS	wind energy conversion system
WFGD	wet flue gas desulfurization
Wh	watt-hour
WLAN	wireless local area network
WMS	waste management system
WNA	World Nuclear Association
WQC	water quality certification

WRAP	Water Reactor Analysis Program
WSCC	Western Systems Coordinating Council
wt	weight
wt/vol	weight per volume
wt/wt	weight ratio
WTE	waste-to-energy
WTP	water treatment plant

X

X	reactance
XIWT	cross-industry working team
XLPE	cross-linked polyethylene
XTP	express transfer protocol

Y

YAG	yttrium aluminum garnet
yd	yard
YIG	yttrium iron garnet
y	year

Z

Z	ceramic
ZCAV	zone constant angular velocity
ZEG	zero energy growth
ZETR	zero energy thermal reactor
ZEV	zero emission vehicle
ZPPR	zero power physics reactor
ZPR	zero power reactor
ZTO	zero time outage

Appendix

Unit of Measure

kilowatt (kW) = 1,000 (one thousand) watts

megawatt (MW) = 1,000,000 (one million) watts

gigawatt (GW) = 1,000,000,000 (one billion) watts

terawatt (TW) = 1,000,000,000,000 (one trillion) watts

gigawatt (GW) = 1,000,000 (one million) kilowatts

thousand gigawatts = 1,000,000,000 (one billion) kilowatts

kilowatt-hours (kWh) = 1,000 (one thousand) watt-hours

megawatt-hours (MWh) = 1,000,000 (one million) watt-hours

gigawatt-hours (GWh) = 1,000,000,000 (one billion) watt-hours

terawatt-hours (TWh) = 1,000,000,000,000 (one trillion) watt-hours

gigawatt-hours (GWh) = 1,000,000 (one million) kilowatt-hours

thousand gigawatt-hours = 1,000,000,000 (one billion) kilowatt-hours

U.S. dollar = 1,000 (one thousand) mills

U.S. cent = 10 (ten) mills

Appendix

Industry Contacts

Alliance to Save Energy (ASE)
1200 18th St. NW Ste. 900
Washington, D.C. 20036
Phone: (202) 857-0666
Fax: (202) 331-9588
Web site: www.ase.org

American Gas Association (AGA)
400 N. Capitol St. NW Ste. 450
Washington, D.C. 20001
Phone: (202) 824-7000
Web site: www.aga.org

American Nuclear Society (ANS)
555 N. Kensington Ave.
La Grange Park, IL 60525
Phone: (708) 352-6611
Fax: (708) 352-0499
Web site: www.ans.org

American Petroleum Institute (API)
1220 L St. NW
Washington, D.C. 20005
Phone: (202) 682-8000
Fax: (202) 682-8115
Web site: www.api.org

American Public Gas Association (APGA)
201 Massachusetts Ave. NE Ste. C-4
Washington, D.C. 20002
Phone: (202) 464-2742
Toll-Free: 1-800-927-4204
Fax: (202) 464-0246
Web site: www.apga.org

American Public Power Association (APPA)
2301 M. St., NW
Washington, D.C. 20037
Phone: (202) 467-2900
Fax: (202) 467-2910
Web site: www.appanet.org

American Solar Energy Society (ASES)
2400 Central Ave., Ste. A
Boulder, CO 80301
Phone: (303) 443-3130
Fax: (303) 443-3212
Web site: www.ases.org
General e-mail: ases@ases.org

American Wind Energy Association (AWEA)
1101 14th St. NW 12th Floor
Washington, D.C. 20005
Phone: (202) 383-2500
Fax: (202) 383-2505
Web site: www.awea.org
General e-mail: windmail@awea.org

Association of Energy Engineers (AEE)
4025 Pleasantdale Rd., Ste. 420
Atlanta, GA 30340
Phone: (404) 447-5083
Fax: (202) 446-3969
Web site: www.aeecenter.org
General e-mail: aeecenter.org

Biomass Energy Research Association (BERA)
901 D St SW Ste. 100
Washington, D.C. 20024
Phone: (847) 381-6320
Toll-Free: 1-800-247-1755
Fax: (847) 382-5595
Web site: www.beral.org
General e-mail: beral@excite.com

Bonneville Power Administration (BPA)
P.O. Box 3621
Portland, OR 97208
Phone: (503) 230-3000
Toll-Free: 1-800-282-3713
Fax: (503) 230-3285
Web site: www.bpa.gov

Consumer Energy Council of America (CECA)
2000 L St. NW Ste 802
Washington, D.C. 20036
Phone: (202) 659-0404
Fax: (202) 659-0407
Web site: www.cecarf.org
General e-mail: outreach@cecarf.org

Council of Petroleum Accountant Societies (COPAS)
3900 E. Mexico Ave. Ste. 602
Denver, CO 80210
Phone: (303) 300-1131
Toll-Free: 1-877-992-6727
Fax: (303) 300-3733
Web site: www.copas.org

Council on Environmental Quality (CEQ)
722 Jackson Place NW
Washington, D.C. 20503
Phone: (202) 395-5750
Fax: (202) 456-6546
Web site: www.whitehouse.gov/ceq

Department of Energy (DOE)
1000 Independence Ave. SW
Washington, D.C. 20585
Phone: (202) 586-5575
Toll-Free: 1-800-Dial-DOE
Fax: (202) 586-4403
Web site: www.energy.gov

Domestic Petroleum Council (DPC)
101 Constitution Ave. NW Ste. 800
Washington, D.C. 20001-2133
Phone: (202) 742-4300
Web site: www.dpcusa.org
General e-mail: info@dpcusa.org

Edison Electric Institute (EEI)
701 Pennsylvania Ave., NW
Washington, D.C. 20004
Phone: (202) 508-5000
Fax: (202) 508-5794
Web site: www.eei.org

Electric Generation Association Power Coalition of America (EGA/PCA)
2101 L St. NW
Washington, D.C. 20037
Phone: (202) 965-1134
Fax: (202) 965-1139

Electric Power Generation Association (EPGA)
800 N. 3rd St.
Harrisburg, PA 17102
Phone: (717) 909-EPGA
Fax: (717) 909-1941
Web site: www.epga.org
General e-mail: info@epga.org

Electric Power Research Institute (EPRI)
3420 Hillview Ave.
Palo Alto, CA 94304
Phone: (415) 855-2000
Fax: (415) 855-2954
Web site: www.epri.com

Electric Power Supply Association (EPSA)
1401 New York Ave. NW 11th Floor
Washington, D.C. 20005-2110
Phone: (202) 628-8200
Fax: (202) 628-8260
Web site: www.epsa.org
General e-mail: epsainfo@epsa.org

Electrical Generation Systems Association (EGSA)
1650 S. Dixie Hwy Ste. 500
Boca Raton, FL 33432-7462
Phone: (561) 750-5575
Fax: (561) 395-8557
Web site: www.egsa.org

Electricity Consumers Resource Council (ECRC)
1333 H St. NW
West Tower 8th FL
Washington, D.C. 20005
Phone: (202) 682-1390
Fax: (202) 289-6370
Web site: www.elcon.org
General e-mail: elcon@elcon.org

Energy Information Administration (EIA)
1000 Independence Ave. SW EI 30
Washington, D.C. 20585
Phone: (202) 586-8800
Web site: www.eie.doe.gov
General e-mail: infoctr@eia.doe.gov

Environmental Protection Agency (EPA)
1200 Pennsylvania Ave. NW
Ariel Rios Bldg
Washington, D.C. 20460
Phone: (202) 272-0167
Fax: (202) 260-0279
Web site: www.epa.gov

Federal Energy Regulatory Commission (FERC)
888 First St. NE
Washington, D.C. 20426
Phone: (202) 502-6088
Toll-Free: 1-866-208-3372
Web site: www.ferc.gov

Gas Appliance Manufacturers Association (GAMA)
2107 Wilson Blvd. Ste. 600
Arlington, VA 22201
Phone: (703) 525-7060
Fax: (703) 525-6790
Web site: www.gamanet.org
General e-mail: membership@gamanet.org

Gas Industry Standards Board (GISB)
1100 Louisiana, Ste. 4925
Houston, TX 77002
Phone: (713) 356-0060
Fax: (713) 356-0067
Web site: www.gisb.org
General e-mail: gisb@aol.com

Gas Processors Association (GPA)
6526 E. 60th St.
Tulsa, OK 74145
Phone: (918) 493-3872
Fax: (918) 493-3875
Web site: www.gasprocessors.com
General e-mail: gpa@gasprocessors.com

Gas Technology Institute (GTI)
1700 S. Mount Prospect Rd.
Des Plaines, IL 60018-1804
Phone: (847) 768-0500
Web site: gastechnology.org

Geospatial Information and Technology Association (GITA)
14456 E. Evans Ave.
Aurora, CO 80014
Phone: (303) 337-0513
Fax: (303) 337-1001
Web site: gita.org
General e-mail: info@gita.org

Independent Petroleum Association of America (IPAA)
1201 15th St. NW Ste. 300
Washington, D.C. 20005
Phone: (202) 857-4722
Fax: (202) 857-4799
Web site: www.ipaa.org

Institute of Electrical and Electronic Engineers (IEEE)
3 Park Ave. 17th Floor
New York, NY 10016-5997
Phone: (212) 419-7900
Fax: (212) 752-4929
Web site: www.ieee.org

Institute of Nuclear Power Operations (INPO)
U.S. Dept. of Energy
(Specify Forrestal Bldg or L'Enfant Plaza Bldg)
1000 Independence Ave. SW
Washington, D.C. 20585
Phone: (202) 586-5000
Web site: www.eh.doe.gov/inpo
General e-mail: esh-infocenter@eh.doe.gov

Interstate Natural Gas Association of America (INGAA)
10 G St. NE, Ste. 700
Washington, D.C. 20002
Phone: (202) 216-5900
Fax: (202) 216-0870
Web site: www.ingaa.org

Interstate Oil and Gas Compact Commission (IOGCC)
P.O. Box 53127
Oklahoma City, OK 73152-3127
Phone: (405) 525-3556
Fax: (405) 525-3592
Web site: www.iogcc.state.ok.us
General e-mail: iogcc@iogcc.state.ok.us

Midwest Cogeneration Association (MCA)
P.O. Box 283
Western Springs, IL 60558-0283
Phone: (630) 323-7909
Web site: www.cogeneration.org
General e-mail: info@cogeneration.org

Midwest Energy Association (MEA)
6012 Blue Circle Dr.
Minnetonka, MN 55343-9138
Phone: (952) 832-9915
Fax: (952) 832-9308
Web site: www.midwestenergy.org
General e-mail: update@midwestenergy.org

National Association of Energy Service Companies (NAESCO)
1615 M St. NW Ste. 800
Washington, D.C. 20036
Phone: (202) 822-0950
Fax: (202) 822-0955
Web site: www.naesco.org
General e-mail: info@naesco.org

National Association of Regulatory Utility Commissioners (NARUC)
1101 Vermont NW Ste. 200
Washington, D.C. 20005
Phone: (202) 898-2200
Fax: (202) 898-2213
Web site: www.naruc.org
General e-mail: admin@naruc.org

National Association of State Energy Officials (NASEO)
1414 Prince St. Ste. 200
Alexandria, VA 22314
Phone: (703) 299-8800
Fax: (703) 299-6208
Web site: www.naseo.org
General e-mail: webinquiry@naseo.org

National Drilling Association (NDA)
1101 Danka Way N Ste. 1
St. Petersburg, FL 33716
Phone: (727) 577-5006
Fax: (727) 572-5012
Web site: www.nda4u.com
General e-mail: info@nda4u.com

National Electrical Manufacturers Association (NEMA)
1300 N. 17th St. Ste. 1847
Rosslyn, VA 22209
Phone: (703) 841-3200
Fax: (703) 841-5900
Web site: www.nema.org

National Energy Marketers Association (NEMA)
3333 K St. NW Ste 110
Washington, D.C. 20007
Phone: (202) 333-3288
Fax: (202) 333-3266
Web site: www.energymarketers.com

National Hydropower Association (NHA)
1 Massachusetts Ave. NW Ste 850
Washington, D.C. 20001
Phone: (202) 682-1700
Fax: (202) 682-9478
Web site: www.hydro.org
General e-mail: help@hydro.org

National Mining Association (NMA)
101 Constitution Ave. NW Ste 500 East
Washington, D.C. 20001-2133
Phone: (202) 463-2600
Fax: (202) 463-2666
Web site: www.nma.org

National Petroleum Council (NPC)
1625 K Street NW, Ste. 600
Washington, D.C. 20006
Phone: (202) 393-6100
Fax: (202) 331-8539
Web site: www.npc.org
General e-mail: info@npc.org

National Regulatory Research Institute (NRRI)
1080 Carmack Rd.
Columbus, OH 43210
Phone: (614) 292-9423 or (614) 292-9679
Web site: www.nrri.org

National Rural Electric Cooperative Association (NRECA)
4301 Wilson Blvd.
Arlington, VA 22203
Phone: (703) 907-5500
Fax: (703) 907-5519
Web site: www.nreca.org
General e-mail: nreca@nreca.coop

Natural Gas Supply Association (NGSA)
805 15th St. NW, Ste. 510
Washington, D.C. 20005
Phone: (202) 326-9300
Fax: (202) 326-9330
Web site: www.ngsa.org

North American Electric Reliability Council (NERC)
116-390 Village Blvd
Princeton, NJ 08540-5731
Phone: (609) 452-8060
Fax: (609) 452-9550
Web site: www.nerc.com
General e-mail: info@nerc.com

Nuclear Energy Institute (NEI)
1776 I St. NW Ste. 400
Washington, D.C. 20006-3708
Phone: (202) 739-8000
Fax: (202) 785-4019
Web site: nei.org

Nuclear Regulatory Commission (NRC)
Office of Public Affairs
Washington, DC 20555
Phone: (301) 415-5575
Toll-Free: 1-800-368-5642
Web site: nrc.gov
General e-mail: opa@nrc.gov

Power Marketing Association (PMA)
P.O. Box 3304
Falls Church, VA 22042
Phone: (201) 784-5389
Fax: (201) 767-1928
Web site: www.powermarketers.com
General e-mail: pma@powermarketers.com

Rural Utilities Service (RUS)
Note: Formerly Rural Electrification Administration (REA)
1400 Independence Ave. SW
Washington, D.C. 20250-0747
Phone: (202) 720-9540
Web site: www.usda.gov/rus

Solar Energy Industries Association (SEIA)
805 15th St. NW Ste. 510
Washington, D.C. 20005
Phone: (202) 628-7745
Fax: (202) 628-7779
Web site: www.seia.org
General e-mail: info@seia.org

Southeastern Power Administration (SEPA)
1166 Athens Tech. Rd.
Elberton, GA 30635-6711
Phone: (706) 213-3800
Web site: www.sepa.doe.gov
General e-mail: info@sepa.doe.gov

Southwestern Power Administration (SWPA)
P.O. Box 1619
Tulsa, OK 74101
Phone: (918) 595-6600
Fax: (918) 595-6656
Web site: www.swpa.gov
General e-mail: info@swpa.gov

Tennessee Valley Authority (TVA)
400 W. Summit Hill Dr.
Knoxville, TN 37902-1499
Phone: (865) 632-2101
Web site: www.tva.gov
General e-mail: tvainfo@tva.com

United States Energy Association (USEA)
1300 Pennsylvania Ave. Ste. 550
Washington, D.C. 20004
Phone: (202) 312-1230
Fax: (202) 682-1682
Web site: www.usea.org

Western Area Power Administration (WAPA)
P.O. Box 281213
Lakewood, CO 80228-8213
Phone: (720) 962-7000
Fax: (720) 962-7200
Web site: www.wapa.gov
General e-mail: corpcomm@wapa.gov

World Nuclear Association (WNA)
Carlton House
22a St. James's Square
London SW1Y 4JH
Phone: +44 (0) 20 7451 1520
Fax: +44 (0) 20 7839 1501
Web site: www.world-nuclear.org
General e-mail: wna@world-nuclear.org

Bibliography

Coalition for Affordable and Reliable Energy. "Bill Summary and Status for the 109th Congress," April 18, 2005; www.careenergy.com.

Edison Electric Institute. "Congress Passes Comprehensive Energy Bill, H.R. 6," July 2005; www.eei.org.

————. "Ensuring Sufficient Generation Capacity During the Transition to Competitive Electricity Markets"; November 2001, pp. 12–16. www.eei.org.

Eisenstate, L., M. Rustum, and M. Farinella. "Benefit of Counsel: Success of Wind Energy Requires Immediate Changes," *Electric Light & Power*, March 2005.

Electric Power Research Institute. "Integrated Environmental Control for Multiple Air Pollutants," Environmental Issues Paper, January 2005.

Electric Power Research Institute. "Power Sec Initiative: Cyber Security Vulnerability Assessment,"Environmental Issues Paper, August 13, 2003 p. 1.

Energy Information Administration. *Annual Energy Outlook 2005*. www.eia.doe.gov (accessed September 22, 2005).

————. "The Changing Structure of the Electric Power Industry 2000: An Update," Executive Summary; www.eia.doe.gov (accessed September 23, 2005).

Federal Energy Regulatory Commission. "Commission Supplements Reliability Policy Statement to Affirm Compliance with Revised NERC Standards." press release, docket no. PL04-5-001, February 9, 2005.

Korman, R. "To Diversify—Or Not," *Energy Biz*, March/April 2005, pp. 10–11.

Kray, M. "Get Nuclear," *Energy Biz*, March/April 2005, pp. 70–72.

Institute of Chemistry at The Hebrew University of Jerusalem, "Steinmetz History," April 26, 2003, p. 7, 11. http://chem.ch.huji.ac.il/~eugeniik/history.

National Energy Policy Development Group. *Report of the National Energy Policy Development Group*, May 2001, overview pp. viii–xv.

National Association of Regulatory Utility Commissioners. BPL Report, 2004, pp. 4–5.

North American Electric Reliability Council. "2004 Long-Term Reliability Assessment Report," September 2004, p. 34.

———. "Urgent Action Standard 1200—Cyber Security," August 13, 2003.

Nuclear Information and Resource Service. "The Pebble Bed Modular Reactor," www.nirs.org/factsheets/pbmrfactsheet.html (accessed May 26, 2005).

———. "The Pebble Bed Modular Reactor," www.nirs.org.

Palombo, C. "The Case for Strategic Asset Optimization," *Electric Light & Power*, March 2005.

Rosenberg, M. "Independent Power Awaits Rebound," *Energy Biz*, March/April 2005, pp. 8–9.

———. "M&A Takeoff," *Energy Biz*, March/April 2005, pp. 16–19.

Senia, A. "Dealing with Greenhouse Gas," *Energy Biz*, March/April 2005, pp. 24–28.

Snell, K. "Preparing for Mercury Regulations," *Electric Light & Power*, March 2005.

Stern, G. M. "Merger Momentum," *Energy Biz*, March/April 2005, pp. 20–23.

Stover, M. R. "U.S. Hydropower: 125 Years Old and Full of Energy," *Electric Light & Power*, March 2005, www.elp.com.

Tesla, Nikola. Speech at the Institute of Immigrant Welfare. Hotel Baltimore, New York, May 12, 1938.

U.S. Department of Energy. *Clean Coal Power Initiative Program Facts*, April 2005.

U.S. Department of Energy. Office of Electric Transmission and Distribution. "Blackout," August 2003; www.electricity.doe.gov.

U.S. Department of Energy. Office of Fossil Energy. "New Fuel Cell Projects to Continue Progress to Zero-Emission Energy," July 19, 2004; www.fe.doe.gov/news/techlines/2004/tl_seca_awards072904.

von Jouanne, A. "The Promise of Wave Power," *Energy Biz*, March/April 2005, pp. 73–74.

Index

C

D

T